YEOVILTON

THE HISTORY OF THE ROYAL NAVAL AIR STATION, YEOVILTON 1940–1990

BY

Cdr. P.M. RIPPON, M.Sc., C.Eng., M.I.Mech.E., M.R.Ae.S., R.N.
&
GRAHAM MOTTRAM M.A. (Cantab), C.Eng.

Graham Mottram.

FIRST PUBLISHED JUNE 1990
FOR THE FLEET AIR ARM MUSEUM, YEOVILTON
by FOX & CO, YEOVIL

TO COMMEMORATE THE GOLDEN JUBILEE
OF R.N.A.S. YEOVILTON

ISBN 1 870872 02 9

CONTENTS

FOREWORD

by Rear Admiral A.R. Rawbone CB AFC

(President of the Society of Friends of the Fleet Air Arm Museum)

Ever since Harald Penrose, chief test pilot at Westland, chanced his luck by landing a Hector in a wet field at Yeovilton, the service men and women of HMS Heron have been developing a solid support for operations at sea, resulting in today's complex modern air base. RNAS Yeovilton, commissioned in June 1940, is the oldest air station in the Royal Navy remaining fully operational. Its varied tasks have been characteristic of the Fleet Air Arm over the last 50 years.

Although vital to the Royal Navy's operations, effective control of the Fleet Air Arm had rested with the Royal Air Force after 1918. Transfer of command from the RAF was initiated, after much political wrangling, by the "Inskip" award of 1937, and was brought about just in time for the rapid expansion needed to meet wartime commitments in 1939. Over fifty aircraft carriers were manned during World War II, posing immense problems for supporting services. That these were overcome so effectively reflects great credit on those who manned bases and supply lines. Not often in the limelight, they frequently saw action and invariably had to improvise to ensure continuity of training and essential supplies.

Yeovilton's role has in many disciplines been dominant. During the turmoil of World War II the station trained naval fighter pilots and parented front line squadrons, but it was the development of air interception techniques and the supply of trained Air Direction Officers that had a critical impact on fleet operations. Experience gained during the war years was enhanced by the almost continuous flow of new developments post war. The evolution of night fighter tactics was a logical consequence and Yeovilton became predominantly a night interceptor and night strike base in the early post war years.

The expense of supplying and maintaining highly specialised permanent equipment highlighted the need to concentrate resources at one base for when squadrons were "working up" or disembarked from their nominated aircraft carrier. Thus night fighter squadrons were always parented by Yeovilton when disembarked and, in consequence, their families were allowed to remain in the station's married quarters when their squadrons were at sea. This resulted in much needed family stability for the first time since the war and has also helped to integrate the air station with its local community. It very soon became the norm for all naval air squadrons, each of which was allocated a parent base ashore.

In later years the most crucial period for Yeovilton occurred during the aftermath of the Defence Reviews of the middle 1960s when the government of the day phased out aircraft carriers. A controversial decision because it was not based on the fact that we could not afford them, but on the false assumption that naval organic air power was no longer needed. Only two naval air stations, RNAS Culdrose and RNAS Portland, were considered necessary to support the surviving fleet. Fortunately common sense prevailed. Yeovilton was reprieved and whilst its tasks were radically reorganized its vital importance to the Royal Navy has since been repeatedly illustrated in the development of Commando Helicopter and Sea Harrier operations. The human factor was paramount throughout and Yeovilton's part in manning and storing an effective Falklands Task Force over a weekend in 1982 was proof of what could be achieved with sound experience, determination and effective training.

Yeovilton has weathered the ebb and flow of political and military discourse, and in true Navy and WRNS tradition its Ship's Company has achieved many and varied objectives with single minded application and unfailing good humour.

In tracing the station's history Peter Rippon has faithfully captured an exceptional fifty years of the Fleet Air Arm. Further in depth research by Peter Anderson, Colin Davies and Graham Mottram, together with the many reminiscences supplied by former members of the Fleet Air Arm and WRNS, add life, amusement and authenticity to the story. For all those who have served at the station the feeling of once again being part of the action, at home at Yeovilton, is irresistible.

EARLY DAYS

In the late 1930s the small Somerset village of Ilchester had become just another agricultural community. Its days of great importance as a Roman walled town and later as the county town of Somerset were long behind it. Its economy, never very robust, now depended on local farmers and the light industry of nearby Yeovil. The neighbouring hamlets of Yeovilton, Podimore, Speckington and Bridgehampton were in much the same condition. The local inhabitants were used to aeroplanes flying overhead. The Westland Aircraft Works had been building aeroplanes since the First World War, and test flights were not uncommon.

One day in 1937 the men working in the fields and the ladies standing chatting in the streets were diverted by a large unusual aeroplane, definitely not of Westland's making, which was flying very low and seemingly in trouble. They watched as the aircraft avoided the River Yeo and the main A303 road and made a safe landing in the large field alongside Yeovilton village. The machine was a Handley Page Harrow twin-engined bomber, forced down by engine trouble, and it was the first aeroplane to make use of the flat plain of South Somerset.

When, in 1938, the management of Westland realised that their Yeovil aerodrome was becoming too small for the aircraft they were building, chief test pilot Harald Penrose was instructed to find a new site. Remarkably, Penrose knew nothing of the Harrow incident, but Yeovilton's large fields attracted his attention from the air as they had done the bomber pilot's. A ''forced landing'' in his Hawker Hector quickly demonstrated the field's suitability. Westland's approach to the Air Ministry was approved, but the land in question belonged to the Ecclesiastical Commissioners. They refused to sell, much to the relief of Mr Parsons of Speckington and Mr Wetherall of Yeovilton, who were farming there.

In 1937 control of naval aviation had returned to the Royal Navy with the ''Inskip Award'', ending twenty years of political struggle between the Admiralty and the Air Ministry. The Fleet Air Arm had to create its own organisation and supplement the few airfields to be transferred from the RAF with new ones of its own, before adopting total control of its own operations in May 1939. Yeovilton was one possible new site, and Commander John Heath, then of the Admiralty's Air Division, travelled down to Somerset to survey the area. He agreed with the proposal, and the land was commandeered from the Ecclesiastical Commission by the Admiralty. Work began almost immediately.

In November 1939, Lt. T.W.Harrington, then serving with 758 Squadron at Eastleigh, drove up with some friends to take a look at the first custom built naval air station since the First World War. He had not realised that the area was part of the flood plain of the River Yeo. The winter had been very wet and they were astonished to see that much of the airfield was a huge lagoon.

This group possibly began a fifty-year tradition when

1. The very first one! A Handley Page Harrow after its forced landing in a field, which was later to become the Royal Naval Air Station at Yeovilton.

2. "Where do you want the runway, Guv?" The crew that built Yeovilton airfield, with steam up and ready to go.

they called in to one of Ilchester's pubs for a beer. Seeing their naval uniforms and wings, one of the locals could not help airing a problem which had been troubling him for some weeks past. Was the Navy intending to build a barrier across the river and flood the fields, so that they could operate seaplanes? The visitors all laughed, told him it was a brilliant idea that they would pass on to their bosses, drank up their beer, and left.

For the next few months the fields where cows and horses had grazed were home to dozens of construction workers. Overseeing the work on behalf of the Admiralty was a Major Dillon, and the Senior Construction Engineer was Mr G.Wilson of the Ministry of Public Building and Works. Westland had come to an arrangement with the Admiralty, and the company were allowed to build a hangar for the final assembly of Spitfires and to use the airfield for test flying. They occupied the western end of the airfield on the Pyle Lane site, and were still there in 1985. Nowadays, though, the building is occupied by the Sea Kings of 707 Squadron.

The first naval officer to command the developing airfield was Commander Robert G. Poole, one of the first batch of inter war naval pilots who had learned to fly in 1924. He was joined early in 1940 by a supply officer, Commander R.H.Rump, who, with his wife, lived first at the Sparkford Inn and then at Yeovilton Rectory, home of the Rev. Arthur Bell.

The two officers needed a secretary, and this job was filled by the first Wren at Yeovilton, Peggy Morton. The WRNS were still in the throes of establishing a new organisation and uniform was scarce. Wren Morton had to make do with a hat of WW1 vintage with a bow. Some of the locals thought she belonged to the Salvation Army.

The ''staff'' set up their offices in one end of the old ''Rag Cottages'', at that time the only permanent

3. Rag Cottage, Yeovilton's first "Ship's Office", with Commander Poole, Captain Murray-Smith and Commander Rump outside. It is not certain how many rooms were used as offices, because a Mrs Lucas, who's husband had been the local undertaker and wheelwright until he died, was still living there. It had proved difficult to re-house her because she refused to go, but the building was demolished when the administration block was completed in August 1940. These were the only houses to be demolished when the airfield was built.

buildings within the airfield perimeter, very close to today's main gate. The other end of the building was still occupied by an old lady, who must have been quite mystified by all the activity which suddenly surrounded her. The Navy's cottage consisted of a large kitchen which was occupied by the officers, a single bedroom upstairs used by two male writers, and a small white-washed pantry or coal house without a window or electricity. The latter would just take a small typing table and chair, and there Wren Morton sat. She was quite happy with the variety of work coming her way, and with the feeling that she was helping to set up RNAS Yeovilton. The pay was 24/- (120p) per fortnight, and she lodged with Dr. Sequeira at the Manor House in Ilchester, where she was well looked after.

Being the only woman brought Peggy Morton a special honour. Her officers felt duty bound to institute a swear box. Anyone who used a naughty word that might have been heard by their Wren had to deposit 6d. There is no record of what happened to the considerable sum collected.

Wren Morton was joined by a second Wren, Phyllis Sequeira, the daughter of the local GP and Peggy's landlord. Wren Sequeira was an "immobile" Wren, one of the many girls who signed up to serve in their home areas, and she became Yeovilton's first M/T driver. Her first vehicle was a black Vauxhall saloon. More Wren drivers joined later, the Vauxhall became one of a fleet of Bedford and Fordson vans and several 3-ton lorries, and the Wrens carried out the whole range of varied tasks required of the road transport section of a large air station for the next six years.

4. Black Vauxhall saloon outside Rag Cottages, the first vehicle in the station's M/T pool, normally driven by Phyllis Sequeira, the local doctor's daughter, who signed up as an "immobile" wren.

Cdr. Rump had the enormous responsibility of organising many of these tasks, as the farmers' fields changed slowly into a usable military airfield, which could operate aircraft and accomodate the people who flew and maintained them. His job was complicated even further as the Luftwaffe began its attacks on southern England. In May 1940 RNAS Ford was severely bombed with heavy casualties, and the decision was taken to evacuate its training units. Yeovilton's runways, essential in such a marshy area, were incomplete, but usable, and Cdr. Poole was briefed to accept No.1 Observers School, comprising 750, 751 and 752 Squadrons. Their motley collection of Sharks, Walruses, Proctors and a Vega Gull were the first FAA aeroplanes to be seen in the neighbourhood. Their stay was short. 750 and 752 Squadrons were despatched with their parent unit to Trinidad, and 751 Squadron went to RNAS Arbroath to form No.2 Observers School.

By now the station was ready to receive its ship's company. On the 2nd June a small detachment of ratings arrived from RN Barracks Devonport to guard and defend the airfield. They were issued with obsolete long-barrel Lee-Enfield rifles and bayonets, with forty rounds of ammunition each, and there were two old Vickers water cooled machine guns mounted on the back of a 3-ton lorry. Whenever there was an air raid warning the lorry would race around the airfield. It would have been of little use; every time the guns were elevated the canvas ammunition belts twisted and the guns jammed.

There was no living accomodation so bell tents and marquees were erected in a field opposite the main gate. The ratings slept six to a tent, laying their open hammocks on duckboards, and eating from field kitchens run by CPO Cook Behenna. Defence manouevres were arranged, with the Polish Army providing the opposition. Later they provided some business for the local hospital!

5. CPO Cook Behenna (standing) built Yeovilton's very first, temporary, galley out of a few bricks borrowed from the contractors.

FLAPS

It is an obvious statement that the organization of the Royal Navy is based on its ships, but less obvious that even shore stations are called ships, and are sometimes known as "stone frigates". Naval air stations are no exception.

The critical point in the creation of any new ship is the day of its commissioning ceremony. RNAS Yeovilton was commissioned as HMS HERON on 18 June 1940. Its new commanding officer was Captain H.S.Murray-Smith, who had previously served with distinction on the China station. His new command housed No.1 Naval Air Fighter School, where new pilots came to learn fighter tactics. His new ship had no housing for its captain; Murray-Smith lived in a rented house at Norton-sub-Hamdon.

His senior officers included Cdr. Jan Sears as Cdr.(Flying), Cdr.(E) Cotton, Surg.Cdr. H. M. Langdale, Lt.Cdr. Dickie Roundell as First Lieutenant, Lt. Woodward as Air Traffic Control Officer, and S/Lt. Young as Captain's Secretary. Cdr Poole, his task of construction now completed, departed to take command of RNAS Fayed in Egypt, and was replaced as Executive Officer by Cdr. Bannerman.

Jan Sears, Cdr.(Flying) in 1940/41, possessed a fiery temper and an awesome reputation. He was another "vintage" naval aviator, who brought dignity to an often light-hearted station. One of his rituals was a "Met Flight" in his personal Sea Gladiator every morning at 7.30 a.m. The aeroplane was meticulously maintained, Sears always entered the cockpit via a set specially made steps, and his pre-flight routine of straps, helmet and gloves was carried out in exactly the same fashion morning after morning. Woe betide any officer or air mechanic who dared to disturb Sears' daily routine. One day a new mechanic joined the ground crew. All went well to begin with, Sears was strapped into his seat, his helmet was fastened and plugged into the radio as it always was, and then an imperious hand reached out to

6. German target map of Yeovilton, dated February 1941. The accompanying information sheet describes it as an aerodrome suitable for fighter squadrons, with 4 hangars, 1 shed, 40 marquees and bell tents, searchlight and AA gun emplacements, and 4 runways of 1100, 900 and 800 metres. Doubtless by 1941 the defences had improved, but early raiders were hardly troubled by the two Vickers machine guns mounted on the back of a truck.

Living quarters and adjacent buildings were estimated at 12000 square metres.

Later on the Germans added Charlton Horethorne to the list of nearby airfields, correctly describing it as an 'action station'.

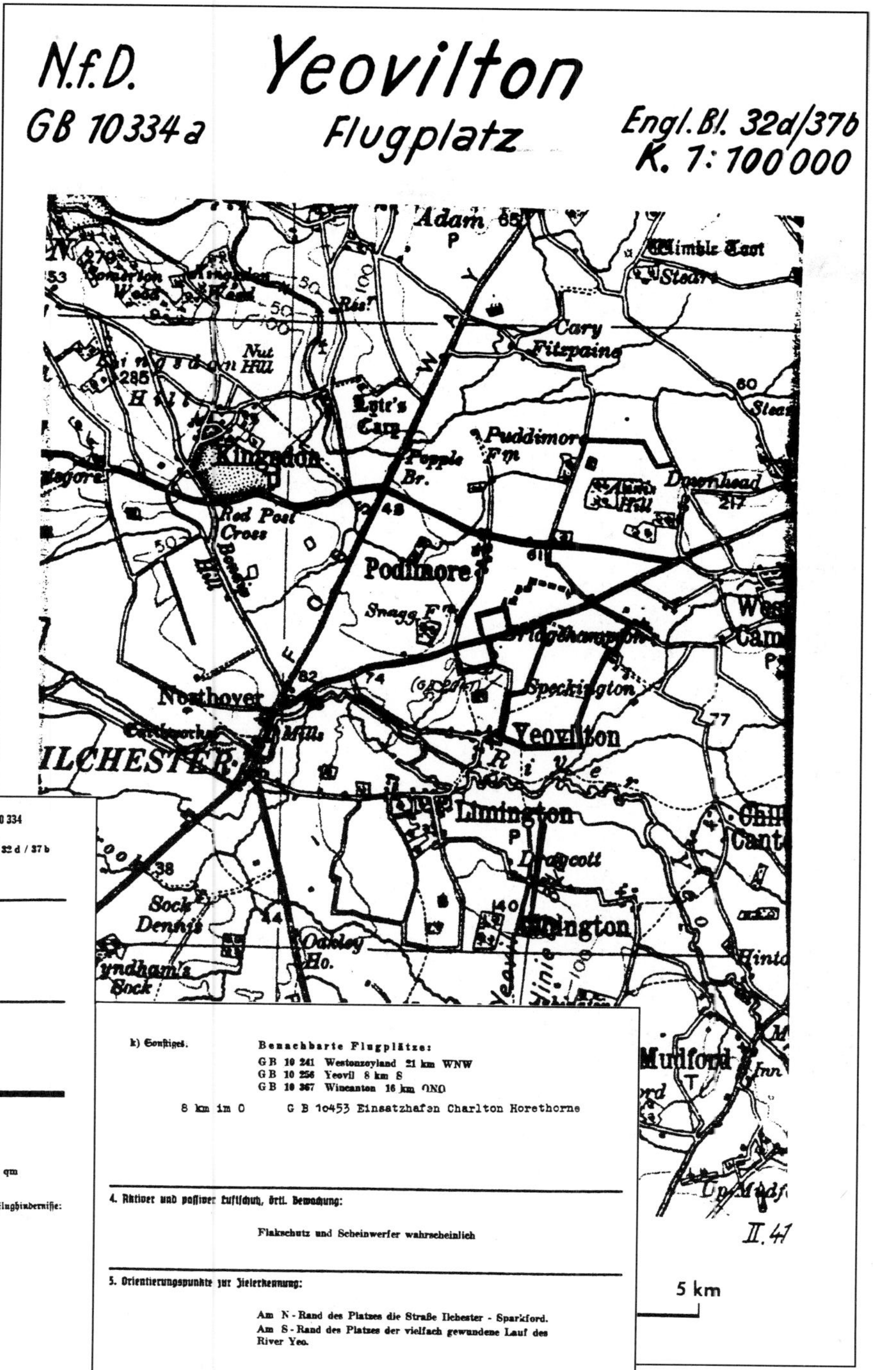

N. f. D.
Zielstammkarte (L)
Land: Großbritannien
England (Somerset)

Ort: Yeovilton
(Nähere Lage) 8 km N Yeovil
Geogr. Werte: 2° 38' W 51° 01' N
Zielhöhe ü. M.: 21 m

Ziel-Nr.: G B 10 334
Kartenblatt-Nr.: Engl. 32 d / 37 b
1 : 100 000
E. B. Nr.:
1 : 63 360

1. Bezeichnung des Zieles:

Flugplatz (mit Luftpark G B 20 41)

Vergl. mit Ziel-Nr.
G B 11 29 Yeovil (Barackenlager) 7 km im SSW
G B 49 43 Somerset (Funkstation) 8 km im WNW
G B 71 60 Yeovil (Flugzeugfabrik) 8 km im S

2. Bedeutung:

Fliegerhorst, für Kampfverbände geeignet

3. Beschreibung des Zieles:

a) Verkehrsanschlüsse: Bahnhof Yeovil 8 km S
Bahnhof Sparkford 6 km NO

b) Ausdehnung insgesamt: Rollfeld etwa 800 x 1100 m Bebaute Fläche: etwa 20 700 qm

c) Art der Anlagen u. Einrichtungen, Bauweise, Bauausführung, Luftempfindlichkeit, Brandgefahr, Flughindernisse:

Teil I:
4 Hallen etwa 8 000 qm
1 Schuppen etwa 200 qm
40 Haus- und Rundzelte etwa 500 qm
Scheinwerferstellung
Flakstellungen
Abstellplätze für Flugzeuge
4 Startbahnen mit 1100 m, 900 m und 800 m Länge

Teil II: Unterkunfts- und Nebengebäude etwa 12 000 qm

Geschlossene Bauweise

Einsturz- und Brandgefahr

d) Erzeugnisse:

k) Sonstiges.
Benachbarte Flugplätze:
G B 10 241 Westonzoyland 21 km WNW
G B 10 256 Yeovil 8 km S
G B 10 367 Wincanton 16 km ONO

8 km im O G B 10453 Einsatzhafen Charlton Horethorne

4. Aktiver und passiver Luftschutz, örtl. Bewachung:

Flakschutz und Scheinwerfer wahrscheinlich

5. Orientierungspunkte zur Zielerkennung:

Am N-Rand des Platzes die Straße Ilchester - Sparkford.
Am S-Rand des Platzes der vielfach gewundene Lauf des River Yeo.

6. Bild- und Kartenunterlagen vom Ziel und vom Zielraum:

a) Anliegend:
a) Kartenausschnitt

b) Außerdem vom Zielraum vorhanden:

7. Gloster Sea Gladiator. Regarded as a pilot's plane and a joy to fly, it was Cdr. Jan Sears' first choice for his early morning 'met' flights. Another Yeovilton pilot took off in a Sea Gladiator in hot pusuit of some German raiders, but was too slow to make contact. Later on he was told by the RAF to 'go and play somewhere else.'

the new rigger for his flying gloves. The unfortunate mechanic shook his Commander's hand firmly and wished him an emotional "Good Luck". Sears "took off" without his aeroplane that morning.

The Germans never recognized a "Battle of Britain" as such. They regarded the events of summer 1940 as a series of air attacks intended to provide the air superiority which had facilitated their invasion of mainland Europe. The official British starting date for the battle is 10 July 1940, but German bombs had fallen on Somerset for several days prior to that date. The Luftwaffe had target maps for both Yeovilton and Westland's Yeovil airfield and both soon knew that they were not to be left alone.

The first attack on Yeovilton came in the late afternoon of 5 July 1940 when an unidentified German dive-bombed the station. Four bombs cratered the airfield, but no-one was hurt. Ten days later a small formation of Junkers Ju88 bombers mounted a deliberate attack on Westland, and one of the bombers split off to make a separate strike against Yeovilton.

Bad weather had prevented RAF fighters from intercepting the raid, and the German pilot used this to his own advantage. He surprised the station by emerging from the cloud base for just long enough to drop his bombs and then make his escape. The gable end of the hangar used by Westland was wrecked, and other buildings were damaged by blast and shrapnel. One of the airfield roads was badly damaged, and there were several more craters across the airfield. Five people were injured, including two men who slid off the hangar roof they were camouflaging.

In August 1940 Captain Murray-Smith moved into the newly-built office block, and the main ship's company and a contingent of Wrens arrived. Many of them came from leading Somerset families, and offered hospitality to the young pilots as a break from their training routine. Their behaviour was closely watched by their "mother hen", Second Officer Waite WRNS, a veteran Wren from the first war.

The new arrivals were left in no doubt that although they might be in rural Somerset, they were very much part of the war. The afternoon of 14 August saw a major air battle all over the county, as Luftwaffe units struck at targets such as RAF Colerne, RAF Middle Wallop and RNAS Yeovilton. Again a single German bomber attacked Yeovilton, this time with no casualties, but its bombs damaged the edge of a runway and aircraft hardstanding, and the petrol bowser which had been parked there.

Understandably, everyone in Britain was nervous, and invasion scares were rife. The nearest army camp was at Houndstone, some seven miles away, and the Colonel was worried about the loss of communications between himself and Yeovilton, should German parachutists attack the airfield. His fall back suggestion was that despatch riders on horses should gallop across country to Houndstone, to call out the soldiers. The Colonel offered to provide the horsemen, because "the Navy probably had no-one who could ride.'' Capt. Murray-Smith must have taken great pride in accepting the offer, with the qualification, ''Actually, we have two Grand National winners in the wardroom.'' Lt. Bob Everett RNVR had won the Grand National in 1929 on Gregalach, and Lt. Frank Furlong RNVR won in 1935 on Reynoldstown.

One of the last invasion scares came in September 1940, just before the Luftwaffe daylight attacks petered out. Late one night Cdr. Rump, still living at the Sparkford Inn, was woken by a messenger from the air station, with instructions to return at once. Dressing quickly, arming himself with his revolver, and with a last farewell to his wife, ''If I don't come back darling, it will be because I am fighting for you and the children,'' Rump set off to walk the three miles back to Yeovilton. The Home Guard, trained by his brother in law, had also been called out, and they gave Rump a rousing cheer as he passed. When he arrived at his post, all was quiet, but parachutists and gliders were expected at any moment. By 6 a.m. all was still quiet and Rump returned home for breakfast. The emergency was over, and there had been no need for the previous night's farewell.

Despite such distractions, HERON was settling into its working routine, and was a ''happy ship''. The ship's company worked hard and mutual respect soon developed between them and their officers. The local population welcomed the newcomers to their shops, home and pubs, and many lasting friendships were sealed over a 4d pint of scrumpy and a game of skittles. Ilchester publicans were particularly pleased when the Pay Office was established in the town, at Ivel House.

The honour of being the first squadron to form at Yeovilton, on 1 August 1940, went to 794 Squadron, commanded by Bob Everett. His job was to tow aerial targets behind Blackburn Rocs and Fairey Swordfish, so that the fighter trainees could practice their aerial gunnery. The novices arrived on 16 September when 759 (Fleet Fighter School) and 760 (Fighter Pool) Squadrons came from Eastleigh. One of the instructors, Lt. A.B. Fraser-Harris, was a survivor of the Norwegian campaign, where he had been shot down and, with his telegraphist air gunner, had trudged through the mountains back to the allied lines. He was one of the first to land at Yeovilton piloting a Sea Gladiator. He was relieved to escape from Eastleigh where, when they were not preparing to repel

8. A flight of Blackburn Rocs. An impressive photograph of an unimpressive aircraft, which was superceded after only a few months. A great deal could be said about its inadequacies, but suffice it to say that it never flew from aircraft carriers, and ended its career as a target towing aircraft. The best part of it was the Boulton-Paul four Browning gun turret, which made the aircraft useful in ground defence, i.e. picketed to the ground with the turret manned by a Telegraphist Air Gunner or Armourer during air raids.

9. Miles Master advanced trainer taxies away from one of the flight dispersals, probably 759 or 760 Squadron, Yeovilton c. 1940/42

10. Blackburn Skua, The Navy's first operational monoplane. The design specification was as a Dive Bomber, but the Navy took it over as a Fighter/Dive Bomber. Unlike its close relative, the Roc, it had some success in both roles, when 800 and 803 Squadrons dive bombed and sank the German cruiser KONIGSBERG at Bergen, during the invasion of Norway, and when one of 803 Squadron's aircraft shot down a Dornier DO18 Flying Boat on 25th September 1939. This was claimed as the first enemy aircraft to be shot down by the Fleet Air Arm. Skuas remained operational until 1941.

invaders, they were sheltering from Luftwaffe bombs. Yeovilton represented rural peace and quiet.

Somerset labourers nearly ended the career of S/Lt. Eric ''Winkle'' Brown RNVR, whose memory of Yeovilton in the autumn of 1940 is of three half-completed runways in a sea of mud. The workmen paid little heed to aircraft lining up to take off, and had to be cleared before every movement. On his first solo in a Blackburn Roc, Brown received clearance to take off, accelerated down the runway and was just getting airborne when he felt a thump on one undercarriage leg. He had hit a pile of rubble dumped on the runway. Brown completed the flight without mishap but, on coming in to land, one undercarriage leg would not come down. He had no alternative but to belly land the Roc in the mud. Although the builders were censured for the rubble, privately the pilots were delighted that there was one damned Roc less for them to fly!

The pattern of training established in those early months continued with only detail changes throughout the Second World War. Pilots arrived at HERON having completed their initial and advanced flying training courses. At Yeovilton they were taught to fight in operational types such as the Sea Gladiator and Skua, and later in Fulmars, Hurricanes and Seafires. Training began with dual in the Miles Master, before going solo on the Sea Gladiator . Dual and then solo formation flying was followed by individual and then section attacks on target aircraft, with camera gun film used to assess gunnery skills. A naval fighter pilot has to operate from a carrier deck, and Aerodrome Dummy Deck Landings (ADDLs) were a crucial part of the training process. Interception exercises, dog fighting, live air firing at ground and towed sleeve targets and dive bombing practice completed the novices' instruction with 759 Squadron. The fully trained pilot was then posted to 760 Squadron where he would continue to build up his hours on service aeroplanes until he was posted to an operational unit.

The Germans' daylight raids on Britain faded away in the autumn of 1940, their invasion plans in ruins. Yeovil and other surrounding towns had been bombed on several occasions, but Yeovilton had escaped any more direct attacks. The Luftwaffe changed its tactics to the nighttime blitz, intending to destroy the morale of British citizens. Yeovilton presented an opportunity target for bombers unable to find Bristol, and alerts became commonplace. The station's personnel spent many nights in the newly-built air raid shelters, but only one raider found the airfield. In the early hours of 12 May 1941 a Junkers Ju88 streaked across the airfield at low level, dropping eight high explosive and eight incendiary bombs. No-one was hurt but the groundsmen had a busy time, repairing the craters in the grassed areas they had sown only days before.

A decoy airfield was set up at Knole on Kingsmoor, four mile to the west of Yeovilton. The area had a similar pattern of roads and river, and false lights and markings were added to complete the deception. The decoy worked on one occasion around the turn of 1941/2. Yeovilton's Armaments and Bomb Disposal Officer, Lt. Tommy Harrington, was ordered to take his team to Knole to defuse four large parachute mines which had been dropped on the decoy. The mines were sticking out of the boggy ground like ancient tombstones, and Harrington left his invaluable Chief Ordnance Artificer at the safe end of their telephone link while he walked out to examine the mines. He soon saw that the safety caps were still in place on the fuses, and decided to have a little fun. He gave one of the mines a hearty kick. The Chief nearly had kittens on the spot. When everyone's nerves had recovered Harrington salvaged the camouflaged parachutes and had them made up into silk dressing gowns for the Chief and his team.

11. A Sea Hurricane of 759 Squadron at its Yeovilton Dispersal in 1943. The squadron's first Hurricanes were transferred straight from the RAF, and were only later converted to the naval version.

12. Aerodrome Dummy Deck Landings (ADDLs). The batsman has just signalled to the pilot that he is in the correct landing position, and to cut his engine power.

The aircraft is a Fairey Fulmar two seat, carrier borne fighter, the first Fleet Air Arm fighter with eight guns. It first entered front line service with 808 Squadron in June 1940, but 759 Squadron, the 'Fleet Fighter School', had them at Yeovilton in May.

13. Grumman Martlet I (AL257), ready for take off. The Fleet Air Arm received its first Martlet Is in the summer of 1940. They were American aircraft and were diverted from a French order after the German occupation of France. The cockpits still had French instrument markings. 759 Squadron, the Fleet Fighter School, and 760 Squadron Fleet Fighter Pool, both used Martlets at Yeovilton from 1940 onwards. When it arrived it was the fastest aircraft in the Fleet Air Arm, very manoeuvrable, and its good deck landing habits made it popular with its pilots. Later marks were named 'Wildcat', in line with the US Navy's designation. A total of 1,093 Martlets/ Wildcats were delivered to the Fleet Air Arm, and used all over the world.

DOWN TO WORK

In the spring of 1941 759 Squadron received about a dozen Hurricanes, all of them well-used veterans released by the RAF. Their skins bore a rash of fabric patches, and these were rapidly added to by HERON's rigours. Their brakes were almost non existent, and their flogged engines leaked oil all over the windscreens. The Serviceability Board at the squadron dispersal often showed "ICL", for Internal Coolant Leak. But for the trainees who had spent most of their time in ponderous Fulmars or outdated Gladiators these old wrecks were a joy. The punch in the back on take off was invigorating, the view for landing was excellent, and the wide track undercarriage was forgiving in a misjudged landing. Once down on Yeovilton's concrete runways, though, there was little friction to aid deceleration, and the brakes had to be used harshly. Little wonder that they often overheated, gave up the ghost and deposited the Hurricane off the end of the runway.

The little airfield at Haldon near Exeter was used for air firing practice, and it was a tight fit for the Hurricanes, with little room to spare for take off or landing. A Deck Landing Control Officer or "batsman" had to be used to bring the fighters in, just as in the limited space of a carrier deck. The target towing aircraft were now Blackburn Skuas plus some Boulton Paul Defiants converted from night fighters.

A new squadron, 787 Fleet Fighter Development Unit, was formed at Yeovilton in March 1941. Its job was to evaluate new fighter aircraft and tactics, and to compare naval fighters against captured examples of the enemy's aircraft. A Fiat CR42 biplane fighter had been captured intact during one of the raids which the Italians mounted in the Battle of Britain, and it was sent to Yeovilton for comparison with the Gladiator. The Italians were still using the CR42 against Fleet Air Arm units in the Mediterranean.

The CO of 759 Squadron's C-Flight, Lt. Rodney Carver, obliged by flying his Gladiator against Lt. "Wiggy" Wigginton in the Fiat. The dogfight over the airfield provided some light entertainment for the "troops" but, despite Carver's operational experience in the Gladiator (he had flown them with 804 Squadron in defence of Scapa Flow in 1940), it was generally mooted that the CR42 won the battle.

The Fiat was also tested against the Fulmar II, its opposition in the Mediterranean, and the Grumman Martlet competed with a German Messerschmitt Bf109E. Although the results of these particular tests have not been located, they could have been of little comfort to the British pilots. Although the Fulmar matched the Fiat for speed, it was much heavier and far less agile than the Italian biplane. The Messerschmitt had already shown that it was stiff competition for Spitfires and Hurricanes, and the Martlet was not in the same class as the RAF fighters. In June 1941 787 Squadron moved to Duxford

14. Boulton-Paul Defiant, converted from a night fighter to a target tug. The Telegraphist Air Gunner in the rear cockpit operates the target winches, which can be seen behind the radiator. Target lines were not all that long, and observers on the ground often marvelled at the nerve of the tug's aircrew.

as the Naval Air Fighting Development Unit (NAFDU), alongside the RAF's AFDU.

For a short time 759 also had a rare bird on strength, a Brewster Buffalo (because a few were in use in the Western Desert), but gradually the squadron standardised on Hurricanes for advanced training, as did its sister unit 760. The latter was commanded by a skilled and popular pilot, Lt. Ken Spurway, another Skua survivor from Norway. His period of command ended tragically when he was killed in a road crash one foggy morning in the autumn of 1941.

Throughout 1941 and on into 1942 the task went on of turning green pilots into men who stood some chance against the enemy. Some of those who qualified joined front line squadrons which were lodging at Yeovilton for a while to reform and work up, while others were sent across the world to join squadrons on carriers or land bases where losses in action had to be made up.

Many of Yeovilton's trainees went on to become outstanding fighter pilots, and some enjoyed distinguished post war careers in the Royal Navy. Many more did not

15. The Navy maintained a healthy interest in enemy aircraft throughout the war. This Messerschmitt ME109F was found abandoned on LG (Landing Ground) 05 at Sidi Barrani, Western Desert. The yellow wavy line indicates that the aircraft belonged to the 9th Staffel of III Gruppe. The naval officer, Lt. L. Phillips RNVR, was from RNAS Dekheila, one of a recovery/technical intelligence party, when 805 and 806 Squadrons occupied LG05 in January 1942.

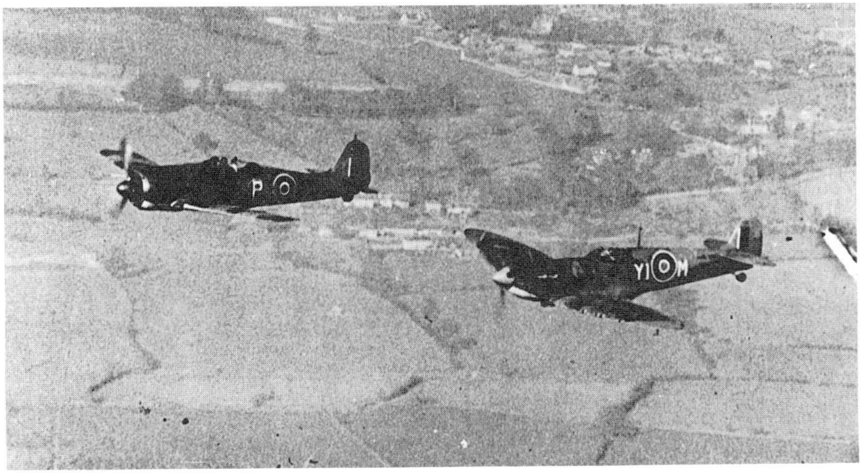

16. A captured Focke-Wulf FW190A of 1426 (Enemy Aircraft Circus) Flight from RAF Collyweston, formates with a Seafire I of 759 Squadron, during a visit to Yeovilton in late 1943. The FW190 is probably FW190A-4/48, W/Nr 7155, captured at RAF West Malling in April 1943, when the German Pilot believed he was landing at a French airfield.

survive to have that chance. The most distinguished of these was a soft-spoken Canadian, Robert Hampton Gray, who was trained at Yeovilton in 1941. By August 1945 he had become Senior Pilot of 1841 Squadron, flying Corsairs from the carrier FORMIDABLE in the last stages of the Pacific War. He had been awarded the DSC for his bravery earlier in the year, and he was probably the most liked member of his squadron.

On 9 August 1945, the day on which the second atom bomb was dropped, he was briefed to lead a sweep over the Japanese coast, and was warned not to take too many chances because the Japanese may surrender at any time. The flight spotted a small flotilla of ships in an inlet called Onagawa Wan, and dived to investigate. The Corsairs flew into an intense flak barrage and Gray elected to bomb an escort ship, the AMAKUSA. His aircraft was hit by a flak shell and burst into flames, but Gray continued his attack and dropped his bomb on target. The blazing Corsair plunged into the waters of Onagawa Wan and was never seen again. Lt. R.H.Gray DSC RNVR was awarded a posthumous Victoria Cross, only the second VC awarded to an Allied fighter pilot in World War II.

17. Lt. Robert Hampton Gray, VC, DSC

In 1941 the Corsair was still years away but the Fleet Air Arm was slowly obtaining improved aircraft, and its naval masters were beginning to realise that naval air power was essential to any Fleet operation. The effectiveness of that power was also improving, thanks to the introduction of Fighter Direction Officers. They were being trained at the Figher Direction School which had been established in a corner of the control tower at the end of 1941.

18. Fighter Direction School, Speckington Manor, 1942/3. Mock-ups of an aircraft carrier's Aircraft Direction Rooms were built on the roof of the house, and in its outbuildings. The telegraph poles around the Manor carry aerials for the transmitters.

"D" SCHOOL

A critical element of victory in the Battle of Britain was the use of radar and fighter controllers, who directed airborne fighters into advantageous positions to intercept German raids. It was essential that the controllers could talk to the fighter leaders by radio telephone (R/T).

The Navy had installed radar in a number of ships but, for decades, the fleet at sea had preferred the comfortable cloak of radio silence. Consequently, early naval air operations had relied on pre-flight briefings, and any signalling was usually from aircraft to ship in Morse by wireless telegraphy (W/T). Gradually the Admiralty accepted their airmen deserved better support, and the concept of the Fighter Direction Officer was developed. The training centre was established at Yeovilton at the end of 1941.

The first commanding officer was Lt. E.D.G. Lewin, (known as ''Drunky'' Lewin). He had won the DSC for his part in the Battle of the River Plate in 1939, when he had flown the Fairey Sea Fox seaplane spotter from the cruiser AJAX. He then converted to fighters, serving for a time aboard ARK ROYAL. Lewin was joined by Instructor Lt. Coke, who proved adept at organising the daily work of the new school, on a shoestring budget and in a corner of the Control Tower.

A section of the airfield was allocated as the practice ground where specially modified Wall's Ice Cream tricycles, the pre-war ''stop me and buy one'' type, were controlled by the trainee Air Direction Radar operators. Some tricycles were ''bombers'' and others were ''fighters''. Both were propelled by sailors or Wrens under the control of the trainees. Each tricycle was fitted with a VHF receiver and headphones for the ''pilot'', and there was a compass by which the pilot followed his or her ordained course. A musician's metronome, calibrated in knots, controlled the pedalling rate, and therefore the speed of the ''aeroplane''. In a high-tech age it all sounds very Heath-Robinson but, apart from the inevitable collisions, the system worked very well. It provided basic instruction in relative speeds and interception courses, without consuming valuable aircraft hours or putting lives at risk.

The effectiveness of this training was demonstrated by EAGLE's first ADR officer, Lt. J. Tricky, RNVR, during ''Operation Harpoon'', a Malta convoy of June 1942. Tricky ensured that his Sea Hurricane pilots knew what to expect of him and soon their interception rate was 100%. More important, these interceptions were taking place well away from the Fleet, and raids were broken up before they presented an immediate threat. ADR officers joined all the carriers and Air Defence Cruisers, and

19. Kenneth More takes a second look at an Air Direction School tricycle, on a visit to the Fleet Air Arm Museum in 1977. These ex Walls Ice Cream machines were manned by ratings or wrens, and played the role of aircraft directed by the trainee Air Direction Officers. The front was built up to restrict forward vision, so that almost total reliance had to be placed on the compass, the metronome which controlled the pedalling rate, and whatever instructions came through the earphones.

fighter effectiveness rose enormously.

The later products of the "D-School", as it became known, were trained in a new schoolroom. By the middle of 1942, as airfield activity built up, it was no longer feasible for the school to lodge in the Control Tower. Speckington Manor, on the north eastern edge of the airfield, was taken over from the army company responsible for airfield defence, and converted to accommodate the school. The upper floors were built to resemble the interior of a carrier's wartop. A separate room held the wireless sets. An RAF party from No.1 Signals Depot, West Drayton arrived in September 1942 to install and commission the new communications gear. The eight RAF tradesmen shared a Nissen hut with a group of seamen who were all recovering from being sunk. The sailors formed the working party, digging cable trenches, and helping to erect GPO telegraph poles which carried the wireless aerials. They also helped the airmen to cope with the naval traditions which still had to be observed even on land, such as saluting the quarterdeck, and tannoy instructions to "darken ship".

By the end of 1943 there were between 60 and 80 officers being trained by 20 instructional staff. After initial practice with the tricycles they worked with real aircraft provided by 790 Squadron, operating Airspeed Oxfords and Fulmars from the satellite airfield, HMS HERON II, at Charlton Horethorne. The three month course was intensive, containing radio and radar theory and practice, navigational and radar plotting, meteorology, and signalling. 759 Squadron's instructors lectured on the mechanics of air fighting to acquaint the trainees with what might be going on in the cockpits of the single seater fighters they would be directing. They practiced gunnery spotting with the army on Salisbury Plain, and with the navy at the RN Gunnery School on Whale Island, Portsmouth. The final week was spent in the trials carrier PRETORIA CASTLE, directing real carrier-borne fighter interceptions.

The newly qualified Fighter Direction Officers were then posted to an operational base, on a carrier, an air defence cruiser, or naval air station anywhere in the world. Many FDOs trained at Yeovilton had distinguished post-war careers. Michael Havers was knighted and became Attorney General. Michael Hordern returned to acting and also became a knight. Kenneth More was by no means unknown as a post-war thespian and his films can still be seen on television. Later in the war Wrens also qualified as FDOs, one of them marrying the boss and becoming Mrs Lewin. Wren FDOs also served overseas at airstations as far afield as Ceylon.

Towards the end of the war the demand for FDOs and the increasing complexity of radar and communications equipment led to Speckington Manor becoming overcrowded. A new purpose built Naval Air Direction Centre was built at

20. Airspeed Oxford AS10, a military version of the Envoy. Served with 790 Squadron from Yeovilton's satellite station at Charlton Horethorne. The squadron was attached to the Fighter Direction School and the Oxford acted the role of enemy bomber.

21. The Canberra T22 started life with the RAF as the PR7 variant, seven of which were converted and delivered to FRADU from 1973 onwards. Its main task was the simulation of anti-ship missile attacks, in which it acted the role of the delivery aircraft, while a Hunter GA11 became the missile. Its Blue Parrot radar is housed in the Buccaneer type nose. The Canberra T22 was replaced in 1985 by the Dassault Falcon 20, and retired to storage at St Athan.

22. The Canberra TT18 was converted from the B2, and entered service with FRADU in 1972, when it took over the target towing role from the Meteor TT20. The high visibility stripes mark it as a target towing aircraft. In this role it can tow either a banner target for aircraft gunnery practice, or the sleeve target used for ships' gunners. However, in this picture the underwing pods are Rushton target winches, with their associated flare throwing targets stowed underneath. When in use the target would be towed some distance behind the aircraft and would act as a heat source simulating target.

23. The 'Blue Herons', FRADU's display team of civilian pilots in Navy aircraft, flying their Hunter GA11s, (upside down on this occasion). Formed in July 1975 the team have disbanded and reformed on several occasions, but are at present 'resting'.

The Hunter GA11 is a conversion of the Mark 4 aircraft, formerly used by the RAF. It entered service with the Royal Navy in 1962, and with the Fleet Requirements Unit at Hurn in 1969. When not 'at play' it is used as a radar target in the training of gunners and fighter controllers. To aid the visual tracking of the aircraft it is fitted with a Harley light in the nose.

Kete in Pembrokeshire and the D-School moved there late in 1945. Fifteen years later the wheel turned full circle and the school moved back to Yeovilton to lodge in the Control Tower once again. The modern radars to be found in HM Ships, types 982/3 and 965 were installed, with intercept computers and all the paraphernalia needed to create the illusion of being aboard a ship at sea.

The school no longer had a dedicated squadron to provide live intercepts, but for some years at Kete the Air Direction Training Unit (ADTU) had been providing support for the trainee Direction Officers, as they were now called, and this unit now moved from St.Davids airfield to Yeovilton. ADTU's aircraft (Sea Venoms) were flown and serviced by a civilian contractor, Airwork Services Ltd. The aircraft were still on military charge but all ADTU's personnel were civilians. Between 1961 and 1972 a hundred and thirty six Directors qualified from the school, including a number of overseas students and the first Wren to become a Direction Officer since the Second World War, Third Officer Jay Taylor.

In addition to ADTU, Airwork Services also provided another second line unit, the Fleet Requirements Unit (FRU), which was based at Hurn Airport, Bournemouth. FRU were the "targets" for naval tracking and gunnery exercises, mostly in the sea area off Portland. As part of the contraction of naval air stations under the "Way Ahead" plan, FRU moved up to Yeovilton in December 1972 and amalgamated with ADTU to form FRADU, the Fleet Requirements and Air Direction Unit. At that point FRADU were the largest "squadron" of the Fleet Air Arm, with 21 Hawker Hunters, 5 English Electric Canberras and 5 Sea Vixens.

FRADU also provided aircraft for the fixed-wing section of the Naval Flying Standards Flight (NFSF). "The Trappers", as they are known throughout the Fleet Air Arm, are responsible for the maintenance of pilot standards, grading all pilots on a regular basis against benchmarks set by the RAF Central Flying School. NFSF also refamiliarise pilots returning to squadron duties from desk or sea appointments, and train Instrument Rating Instructors and Qualified Flying Instructors for service within the squadrons.

FRADU retired their last Sea Vixens in February 1974, replacing them with Canberra T22s. These had Blue Parrot radar in the nose, making them particularly suitable for anti-shipping exercises.

Although all the aircrew flying with FRADU are civilians, they are all very experienced ex-service jet pilots. Their typical age is about 45, and they all have several thousand hours in their logbooks. Their experience was demonstrated in public by their formation of ''The Blue Herons'' aerobatic team in July 1975. They were the first aerobatic team in the world in which civilian pilots flew military jet fighters, and for some time they were the Navy's only fixed wing representatives on the airshow circuit. Their Hawker Hunters gave immaculate performances for five years, winning the Shell Trophy at the 1977 International Air Tattoo at Greenham Common. The Blue Herons disbanded in 1980 when their leader, Derek Morter, retired from flying.

In December 1983 Flight Refuelling/RCA Services took over the FRADU contract from Airwork, and assumed the running of the fleet of 12 Canberras and 24 Hunters. Some of the ageing Canberras were later replaced by Falcon 20 aircraft. A fleet of 16 Falcons was built up, and they took over the threat simulation tasks from the Canberras. The Falcon can also pose as an electronic warfare threat, with two operators in the cabin able to jam communications and radar equipment over a wide frequency range. The Falcons later moved their operating base to Hurn, in another full circle move, taking some of the FRU duties away, whilst leaving the Hunters at Yeovilton, predominantly for Air Direction and NFSF work.

The Air Direction School continues its task today, producing qualified Direction Officers able to control shore-based or Allied forces support aircraft from HM ships. As in all spheres, a sophisticated computerised Aircraft Control Trainer is now in use, which can simulate all possible scenarios and tactical situations in a way that the Wall's Ice Cream tricycles never could. Trainee controllers spend 15 weeks of their 20 week course at Yeovilton, with 80 hours on the simulator, and a further 40 controlling live aircraft sorties flown by FRADU aircraft.

24. Dassault Falcon 20, operated by Flight Refuelling Ltd in the Fleet Requirements role. The majority of these aircraft were formally Federal Express parcel carriers. They replaced the Canberra T22s in the Radar Target role, with the use of various ECM/ESM pods, two of which can be seen here fitted to the underwing pylons. Flight Refuelling's Falcons are also used in the long range communications role to such places as Decimomannu and Norway.

THE LATER WAR YEARS, 1943-45

The Admiralty's realisation of the worth of naval air power and the plans for the invasion of mainland Europe, followed by the final battles in the Far East, led to a massive expansion of the Fleet Air Arm. The training burden on Yeovilton became even greater. Charlton Horethorne was opened as a satellite, HMS HERON II, in 1942, and in April 1943 RNAS Henstridge was commissioned as HMS DIPPER, to spread the load.

The majority of the new aviators, as had been the case for most of the war up to now, were recruited straight from civilian life. There was no time for them to undergo proper naval training and, since the Admiralty still did not regard aviators as proper naval officers, these young men entered the Services as members of the Royal Naval Volunteer Reserve, the ''Wavy Navy''. They enlisted under the ''Y-scheme'' as Acting Temporary Naval Airmen, and were only commissioned after they had obtained their wings. A great many of them were fortunate enough to do their initial flying training in the United States or Canada, where there were lots of pretty girls and no rationing. The social life in blacked-out wartime Somerset was something of a comedown, but the compensations lay in the high performance fighters they now learned to fly.

There were still a few old aeroplanes at Yeovilton, looked after by Station Flight. The Sea Gladiator was such a delightful ''pilot's aeroplane'' that a succession of senior officers ensured that one stayed serviceable at Yeovilton throughout the war, long after its operational life was over. When the burden of flying a desk became unbearable, the Gladiator would be ordered out for an invigorating aerobatic spin, or maybe just a pleasant cruise over the Somerset countryside, away from the telephone. A few other aeroplanes were kept for communications flights, and a variety of types, such as Swordfish, Fulmar and Skua remained in use until they were beyond economical repair.

Apart from the station ''hacks'', the outclassed aircraft of earlier years had gone, replaced by Sea Hurricanes, Wildcats and Seafires. Yeovilton continued its task of converting novices onto high performance fighters, whilst No.2 Naval Air Fighter School at Henstridge became the primary base for type training and squadron work-ups on Spitfires and Seafires. This was particularly important from the end of 1943 onwards, as part of the preparations for the D-day landings. New duties were added to the naval fighter's repertoire, close ground support and spotting the fall of shot for the Fleet's guns during shore bombardments.

Training for experienced fighter pilots was added to Yeovilton's role in May 1943 with the formation of 736 Squadron as the School of Air Combat. Its CO was one of the Fleet Air Arm's most distinguished fighter pilots, Lt.Cdr. R.E. ''Jimmy'' Gardner, whose operational career had begun in Douglas Bader's 242 Squadron in the Battle of Britain. His job was to bring naval fighter pilots up to date with the latest air combat techniques. The squadron moved away in the autumn to become part of the School of Naval Air Warfare at St.Merryn in Cornwall.

If anyone had any doubts about the severity of the forthcoming battle for Europe, Yeovilton personnel could soon dispel them. Towards the end of 1943, as the USAAF bombing campaign built up, several Boeing B-17 Flying Fortresses made emergency landings at HERON. Torn by flak or fighter shells, with severely wounded crewmen aboard, these American bombers demonstrated graphically the strength of German defences. When the dead and dying had been attended to, the large bombers were towed to a dispersal to await the arrival of a USAAF mobile workshop. Sometimes the Americans had to virtually rebuild the Fortress before it was fit to fly out, weeks after its arrival.

On one occasion a Liberator pilot chose Yeovilton for a precautionary landing. The noise of his preliminary

35. Yeovilton closed for flying in September 1945. As it had become one of the main demobilisation centres, there was plenty of spare labour on hand to form a 'Runway Working Party'. Several officers were sent on crash courses in runway construction, and then 'controlled' operations under the guidance of two civilian engineers. There were a few pieces of mechanised plant, as seen here, but most of the work was completed by sheer weight of numbers, helped along by a light hearted mood, because the war was over and they would soon be going home.

tified cause for a massive celebration. Every Very light that could be found was fired off into the Somerset sky and for the *piece de resistanc* every blackout board from HERON's windows was stripped out and burned ceremonially on an enormous bonfire.

After five years of continuous use Yeovilton's runways were in poor condition and, with the end of the Pacific War, the station was closed for flying in September 1945. Charlton Horethorne had been exchanged for Zeals in April 1945, and both "Charlie-H" and Yeovilton squadrons transferred their operations to the new air station, HMS HUMMING BIRD.

Yeovilton had become one of the main demobilization centres, where at any one time hundreds of men awaited their honourable discharge from the Royal Navy or RNVR. This pool of unemployed labour was put to work as the Runway Working Party. A number of officers were sent on crash courses in runway construction, and "controlled" operations under the guidance of two rather frustrated civilian engineers. There were a few pieces of mechanised plant to help, but the job was completed mainly by sheer weight of numbers. Armed only with picks and shovels, the RWP dug up thousands of feet of runway and then laid new foundations with large stones at the bottom and smaller ones as infill. Civilian contractors topped off the foundations with the final layers of tarmac.

By January 1946 759 Squadron was back at Yeovilton, Zeals was closed, and all flying had ceased at Henstridge. HERON's only squadron, 759, was absorbed into 794 Squadron at Eglinton in Northern Ireland in February 1946, and that station, No.3 Naval Air Fighter School, took over the task of No.1 NAFS from Yeovilton. After a short and intensive wartime life, the rundown of Yeovilton seemed dramatic and complete.

36. An early Meteor III stopped off at Yeovilton to let the Navy take a look at it, while on its way to re-equip 616 Squadron, Auxiliary Air Force, in late 1944/early 1945. 616 Squadron at this time had moved to Colerne in Wiltshire, having had some success against the V1 flying bombs with the earlier Meteor I, when the squadron was based at Manston.

THE POST WAR YEARS

After the almost continuous noise of the previous years, local villagers found the peace of early 1946 rather uncanny. Many thought the camp would be closed. It was not, but its role was changed. HMS HERON continued to be a training centre, but now it produced air mechanics and ''plumber pilots'',, the Fleet Air Arm's nickname for Engineering Test Pilots.

A new resident squadron arrived in April 1946. 700 Squadron, the Maintenance Test Pilot Training Squadron, flew in from RAF Middle Wallop. The unit ran ten week courses, and the students were engineer officers who were also top class pilots. They learned to fly all the aircraft types in current service and consequently the squadron had a wide variety of aircraft on strength.

But Yeovilton's main task was the training of hundreds of air mechanics. Four specialisations were catered for, airframes, engines, electrics, and ordnance. All trainees began with basic instruction in materials and use of tools, and throughout their training there were also classes in general education. Airframe mechanics worked in the Airframe Repair Shop, on aircraft such as the Tiger Moth and Seafire. The Dope Shop taught them how to paint aeroplanes, and the Hydraulic Bay introduced them to the complicated plumbing of the undercarriage, and other hydraulic systems.

Engine mechanics learned how to strip and rebuild the aero-engines of the day, and their ancillaries. All their efforts were then run on a test bed to check the quality of the trainees' work. Electricians were given a comprehensive course in all aspects of aircraft electrical systems, and ordnance mechanics learned the safe handling of all armaments then in use, from light machine guns to pyrotechnics and rocket projectiles. Every course finished with a spell in the Servicing and Flight Section, where trainees carried out routine ''daily inspections'' on all types of aircraft. Training in aircraft salvage and firefighting, and the correct documentation of all aircraft maintenance, brought the course work to an end. The written exam followed, and obviously a pass meant a great deal to the young novices.

By 1947 training had been expanded to include Special Air Maintenance Courses (SAMCOs), which were detailed courses dealing with particular aircraft and their systems. Pupils ranged from junior ratings to Air Engineer Officers, and courses began to include some overseas students. The population of HERON had risen to around 2,700 officers, men and Wrens.

The autumn of 1947 saw the first of the very popular Air Days at Yeovilton, initiated by Capt. W.K. Edden and organised by Lt.Cdr. Peter Davis. Air Days had been a

37. Trainee Air Mechanics prepare to start the Merlin engine of an old Seafire MkIII for ground running practice at Yeovilton, 1947.

38. Seafire MkIIIs stripped as Airframe Instructional Aids for trainee Air Mechanics, May 1947.

39. Yeovilton's Wrens dancing a hornpipe at the 1947 Airday.

regular pre-war event at RAF stations, but this was the first chance the Royal Navy had to demonstrate its own post-war aviation activities. The early ones were very much public relations exercises, to develop links with the local community, but as time went on and entrance fees were charged, another main objective was to raise money for naval charities. By today's standards the 1947 event was more of an "Open Day", with the emphasis on displaying all aspects of Yeovilton's normal working routine. Manufacturers who supplied aircraft and equipment to the Fleet Air Arm gave tremendous support with their own stalls and displays to add to the FAA's own efforts. Visitors could inspect examples of the trainees' work, and view some of the living quarters and the galley. There was also a limited flying display from the resident squadrons, and Westland's new helicopters, Dragonfly and Widgeon, plus a flypast of current naval aircraft.

By the end of 1947 courses for Pilot's Mates had begun. The routine of four different mechanics each doing a "daily inspection" was recognised to be wasteful, and a Pilot's Mate would be solely responsible for all elements of a D.I. Specially selected air mechanics went through an intensive course, with a written examination at the end. Only after another three months experience on a squadron would the Pilot's Mate qualify to sign Form A700, the serviceability status of an aeroplane.

The start of work on the new Married Quarters in Ilchester signalled the intent that HERON was to be a permanent peacetime station. Prior to this development, officers had lived in the wardroom, and ratings in the appropriate messes. Very few people lived out in rented accommodation, and most of the messes had a full entertainments programme to fill the weekday evenings. At one point a fishpond was built outside the CPO's Mess by its members. Angling matches were added to the programme, with very strict rules. Lines had to be of black thread, floats had to be matchsticks and the worm bait had to be tied to the end of the thread.

By 1950 Married Quarters were in reasonable supply and "living out" had become established. The great benefit was that families could now be together, near to the husband's workplace. The new quarters were extremely well equipped, and the regular business from the new population was very welcome to Ilchester's traders. At first the quarters were part of "Captain's Rounds", a periodical inspection by the Commanding Officer, but this practice did not survive the protests from navy wives, who complained that "rounds" were an affront to their housekeeping.

700 Maintenance Test Pilot Training Squadron disbanded in September 1949, and their place was taken by 767 Squadron, transferred from Milltown, whose job was to train Deck Landing Control Officers (Batsmen). The pilots had to fly circuits and landings incessantly, and hence the unit had earned its nickname of the "Clockwork Mouse" squadron.

When the Korean War broke out in 1950, one of the Royal Navy's Light Fleet carriers, HMS TRIUMPH, was sent to the war zone immediately. Until the Armistice three years later there was always a British carrier, with its squadrons of Sea Furies and Fireflies operating in Korean waters. The Fleet Air Arm had been reduced substantially at the end of the Second World War, and the demands on manpower made by its involvement in Korea overstretched the peacetime service. A large number of reservists, both aircrew and groundcrew, with wartime experience were called up. Many of these men, particularly the air mechanics, received their refresher training at Yeovilton. Few of them were ever required to serve in the Korean war zone, but their efforts at home stations enabled the Fleet Air Arm to sustain its normal activities until the Korean War ended in 1953.

40. Fairey Firefly NF1, PP617, c.1948. The Canister mounted under the engine contains radar equipment, and distinguishes this aircraft as a night fighter.

41. Hawker Sea Fury FB11, in Korean War markings. 1834 squadron RNVR operated Sea Furies from Yeovilton in 1954/55.

In the autumn of 1951, a fire broke out at the oil tank installations at Avonmouth. A group of 30 volunteers travelled from Yeovilton to assist. The dockside was an inferno. The sailors stayed all through the night, acting as porters for the firemen hosing down the unexploded oil tanks to keep them cool. When daylight came they could see the whole docks area under 18 inches of water as a result. The sailors had earned the fry-up waiting for them when they returned to Yeovilton.

Another change in Yeovilton's role came in 1952, when it was allotted the duties of the All-Weather Jet Fighter Station. The airfield was still essentially a Second World War station; during night flying ''goose neck flares'', little more than cans of burning paraffin, were used to mark the runways. Now the runways had to be extended to handle the faster aircraft, and the support facilities improved to bring the station up to modern NATO standards. So for a while Yeovilton closed down and Henstridge became home to its squadrons. Taylor Woodrow began work in April 1952, and gradually the new hardstandings and runway extensions began to take shape.

On 27 May 1953, the Headquarters of Flag Officer Flying Training (FOFT) moved from Donibristle to Yeovilton, and half of the administration block was taken over by FOFT's staff. With the airfield closed to fixed-wing flying, FOFT's communications helicopter was the only aircraft to use the station for some time.

When Yeovilton opened again after a period of nearly eighteen months, the first unit to move in was 764 Squadron, the Operational Flying School, previously at RNAS Lossiemouth. The squadron was the last one to

42. A Whirlwind HAS22, as used by FOFT as his 'Admiral's Barge'.

43. Vickers Supermarine Seafire 17, Griffon VI engined, first entered service in September 1945. Longest lived of the post-war Seafires, it remained with training and RNVR squadrons until 764 Squadron disbanded at Yeovilton in November 1954.

44. Hawker Sea Hawk FGA6 XE458, built by Armstrong Whitworth. The pylons under each wing are carrying 3" rocket projectiles, 500lb bombs and 100 gallon drop tanks.

operate Seafires (bringing with it 13 Seafire 17s) and the first to operate jet fighters from Yeovilton, with three Hawker Sea Hawks.

In July of 1953, Yeovilton was nominated as the parent station of the RNVR's Southern Air Division, but because the airfield was closed 1834 Squadron, which drew its recruits from the West Country, used RAF Benson in Oxfordshire as its base to reform as a fighter squadron on Sea Furies in October that year. When the squadron moved base to Yeovilton in January 1954 its arrival was spectacular, having an escort of Sea Furies, Seafires and Fireflies drawn from the other RNVR Air Divisions. The current FOFT, Rear Admiral Guy Willoughby, was waiting on the tarmac to greet the unit.

The RNVR Air Divisions were not known as ''weekend aviators'' for nothing. In most respects they were run like a Territorial Army unit, with only a few permanent weekday staff, maintaining aircraft and preparing for the weekend, when the squadron would spring into action with pilots and ground crews practicing their respective skills throughout Saturday and Sunday.

In March 1954 Yeovilton adopted its new role and the first All-Weather Jet Fighter front-line squadron, 890 Squadron, formed up, equipped with de Havilland Sea Venoms. Station and squadrons soon settled in to their new roles, all-weather training and operations from a single base. All crews came direct to Yeovilton for training, joined squadrons at the end, and returned to Yeovilton after deployments to carriers or overseas stations.

Cdr. R. Anderson was appointed Executive Officer in 1957. He had little experience of the Fleet Air Arm, and was quite astonished at the station routine. Compared with the normal shore establishment, with its ''eight to four'' mentality and the silent hours that followed, Yeovilton had a terrific sense of urgency. Aircraft flew all day and most of the night, and ground crews worked all hours to keep their machines on the line.

Anderson also had to cope with the dangers of naval flying. In one six month period the station lost twelve aircrew in accidents, and he always accepted it as his sad duty to inform personally the next of kin living in the neighbourhood.

Although Anderson found his aviators to be exasperating at times, particularly in the mess, he felt privileged that he had been allowed to share the amalgam of esprit de corps, morale, courage, dedication and sheer professional skill which made the Royal Navy's air arm one of the finest in the world.

At this time those qualities had been tested in action when some of Yeovilton's squadrons were called upon to fly in real operations. The ill-advised Suez campaign began with airstrikes against Egyptian airfields. 809 Squadron from BULWARK, 892 and 893 Squadrons from EAGLE took part in the attacks of November 1956, flying Sea Venoms.

The Sea Venom was an important aeroplane in the technological development of the Fleet Air Arm. The first naval mark, FAW20, brought Airborne Interceptor Radar in a jet fighter, and the last, FAW22, introduced guided missiles to the fighter armoury. 893 Squadron was the first unit to receive Firestreak armed Sea Venoms in 1956.

The intense activities of the 1950s, combined with the advent of the much heavier de Havilland Sea Vixen, prompted another upgrading of Yeovilton's runways. At the end of October 1956 the airfield was closed again. The Laing Company worked throughout the whole of 1957 to put a further 8 inches of concrete on the main runway. The opportunity was also taken to reroute the perimeter track. The new track was laid clear of all the obstructions, such as hardstandings and aircraft parks, which had previously made taxying difficult. Although Yeovilton had been closed to flying for the equivalent of three years out of the six between 1952 and 1958, the programme of major improvements had prepared the station for the big jets that were to come.

45. Sea Venom WM506/227 of 809 Squadron, putting the Spray Assisted Arrester Gear (SPRAG) on Yeovilton's runway to its first test in 1954.

46. Sea Venom FAW21 fitted with de Havilland Firestreak infra-red homing air-to-air missile, c.1960. Yeovilton's Sea Venom Squadrons helped develop the Firestreak in preparation for the Sea Vixens, then just entering service.

47. Sea Vixen FAW1 of 766 Squadron with its alternative weaponry. Three different types of early missile are shown; Firestreaks fitted to the inner pylons, Red Tops standing beneath the air intakes, and a pair of Bull Pups on the trailer in the foreground. Outboard of the Red Tops are various forms of 2" rocket, and under the wingtips are four 500lb bombs. In front are sixteen 3" rockets each side. Drop tanks are fitted to the outer wing pylons.

THE SEA VIXEN YEARS

The new aircraft which arrived at Yeovilton in 1958 was a second generation jet, much heavier, more powerful, and much more complex than its predecessors. The de Havilland Sea Vixen was to be a familiar sight around Yeovilton for many years to come.

The Sea Vixen Intensive Flying Trials Unit (IFTU), 700Y Squadron, was formed in November 1958 and spent the next eight months developing the operational and servicing techniques ready for front line use. The new concept of a flying weapons system, with AI18 Radar, Ferranti Pilot Attack Sight, and Firestreak missiles, had to be fully tested, as well as the delivery of the ''old'' weapons such as iron bombs and 2'' rockets.

When its work was completed, the IFTU provided the nucleus of the first front line squadron, 892, which formed on 1 July 1959. Later in the year the squadron detached four aircraft into 892 B-Flight, to concentrate on further development of the Firestreak missile.

Aircrew training for the Vixen was provided by 766 squadron, also at Yeovilton, which received its first aircraft in the autumn of 1959. The All-Weather Fighter Training Course comprised 14 weeks of intensive effort. Crews had to achieve 110 flying hours, of which 40 were at night. Training on these sophisticated new aircraft was supported by the new technology of ground simulators. The Air Intercept Ground Trainer was an integral part of the Observers' Course. It usually operated a ten-hour day, teaching observers how to use their Airborne Intercept Radar to guide the pilot to their target. Pilots also had an aircraft simulator, which reproduced on the ground all the flying characteristics of the Sea Vixen, and reduced the accident risk in the early stages of flying the actual aeroplane.

There was still time for social life, and Mess Dinners were usually arranged on a Friday night, when the station was looked after by a Duty Crew. The late evening of Friday 20 March 1959 was much like any other Friday for the Duty Air Traffic Officer, until a distress call was picked up. An aeroplane had engine trouble and its pilot requested immediate assistance. Air Traffic ''talked'' him into Yeovilton and the pilot made a safe landing in the dark. When they asked him where he came from the orange-clad pilot replied ''Houston, Texas''.

William R. Mullen was Regional Sales Manager for Mooney Aircraft Inc. of Kerrivale, Texas. He had taken off from Houston in the single engined Mooney 20, with the object of breaking the world solo distance record, with a view to the obvious publicity for his product. Mullen had flown a ''great circle'' route over New York, Newfoundland and the North Atlantic, expecting to make a landfall around the Bay of Biscay. He had failed to beat the record by about 50 miles, but he had flown nearly 5,000 miles in 32 hours and 7 minutes. Despite his garish flying suit, Mullen was invited to join the Mess Dinner, where he made a triumphal entry shortly after the Loyal Toast. Mullen's speech of thanks included the information that he was trying to make Paris, but his engine seemed to want to visit Yeovilton.

The intensive day and night flying programme imposed a heavy strain upon ground crews. 766 Squadron introduced a system of continuous servicing cover, with personnel divided into three watches, so that aircraft could be turned round rapidly between flights, at all times of the day and night. The monthly utilisation of aircraft improved by 40%, one aircraft flew for 575 hours in one month, and 400 hours per month was the average. In the words of a veteran from the period, ''The squadron was flying twenty-four hours a day and it was bloody hard work for everyone''.

One of the longest-lived Sea Vixen squadrons was 899, which reformed in February 1961 as the Sea Vixen FAW1 Headquarters Squadron. Its main tasks were to conduct tactical and operational trials with the aeroplane, including ground handling and maintenance, leading to the compilation of the Naval Air Tactics Notes. In-flight refuelling was one of the unit's specialities, providing disembarked squadrons with refuelling practice. Some pilots needed it! Various characters speared the refuelling drogue with a Firestreak missile or the end of a drop tank, instead of the probe intended for the purpose. One observer was particularly disconcerted to hear and feel the tanker's drogue banging on the hatch above his head!

In December 1961 the Mayor of Yeovil, Alderman C.E. Hawes JP, was presented with a silver Heron inscribed, ''From the Captain and People of HMS HERON, RNAS Yeovilton, to the Mayor and People of

48. 'Buddy-buddy'. In flight refuelling of one Sea Vixen FAW1 of 899 Squadron by another, in the early 1960s. The refuelling hose was carried in a special Flight Refuelling pod, mounted on the starboard underwing pylon of the supplying aircraft.

49. A Sea Vixen FAW1 of 899 Squadron being prepared for flight. The ground starter is plugged in ready to start the engines, the Armourers are checking firing circuits, supervised by the PO leaning over the cockpit, while an Air Mechanic (lying on his back) is locking the Dive Brake, which had a tendency to fall on ground crew unexpectedly, back into its flight position.

Yeovil to commemorate 21 years of Friendship and Goodwill in War and Peace - 1961''. The compliment was returned by the Mayor and council conferring the Freedom of the Borough upon the station at a ceremony at Yeovil Town Football Ground on 16 June 1962. Captain Rodney Carver, the 1940 instructor and now CO of Yeovilton, accepted the Certificate of Admission to the Honorary Freedom of the Borough, contained in a silver casket which also held the key of the town, and signed the Freemen's Roll. The casket was trooped through the paraded ranks while a slow fly-past proceeded overhead. Parades through the streets and a Civic Luncheon completed a proud day for the station and its adoptive town.

During the summer and autumn of 1962 the instructors of 766 Squadron formed an aerobatic display team known as ''Fred's Five'', led by Lt. Cdr. P.B. Reynolds. Alliteration held as much allure for them as aerobatics, with their catchphrase of ''Fabulous, Fearnought, Frocked Fred's Five Frequently Flabbergast Farflung Fraternities

50. Yeovilton greets Yeovil. Capt. W. Simpson, CO of Yeovilton, presenting a silver heron to the Mayor and Mayoress of Yeovil, Alderman and Mrs Kelly, at HMS HERON, in December 1961, commemorating twenty one years of friendship between the air station and the town. This splendid piece of silverware is presently nesting in the Mayor's Parlour in the Town House.

Flying Fiendishly For Free (Firkins Favoured)''. Their aerial performance was no less breathtaking, and their part in the SBAC show at Farnborough, joined by the Vixens of 893 and 899 Squadrons, and the Scimitars of 736 Squadron from RNAS Lossiemouth, left no watcher in doubt about the airmanship of the Fleet Air Arm.

South Somerset just missed a white Christmas in 1962, but the snow began to fall on Boxing Day. Once started it kept falling, and there was soon a thick layer all over the West Country. Yeovilton's runways quickly became unusable, but with most of the station away on leave, and the well known fact that ''snow does not lie for long in Somerset'', nothing was done except to wait for the thaw. When that came it lasted just long enough for the subsequent freeze-up to turn the snow into hard packed ice which nothing mechanical would shift. In desperation, two old Nene jet engines were mounted on the back of a lorry and driven carefully down the runways so that the hot jet gases could melt the ice. The resulting sheets of water promptly froze again into sheets of glassy-smooth ice. The only way to clear the runway was by pick and shovel. Working parties were formed up and they struggled for days on end to shift the ice sheets. By the end of January only one runway was back in action. The big freeze lasted until late March of 1963.

Throughout the 1960s Britain's Nuclear Deterrent depended on the RAF's V-bomber force. The massive Vulcan, Valiant and Victor bombers were easy targets for an enemy's pre-emptive strike, and so the bomber squadrons practiced rapid dispersals to airfields all over the country, including Yeovilton. The V-bomber Dispersal was on the south side of the airfield, where the RAF teams had their own messing arrangements. Every deployment was secret until a few minutes before the bombers' arrival. ''Secret'' can be a word of limited

51. Instructors of 766 Squadron, led by Lt. Cdr. Peter 'Fred' Reynolds, piloting Sea Vixen FAW1s, practice as the display team, 'Fred's Five', in 1962.

They had to dovetail their display team practice into the normal work of the Squadron, and this entry in the Squadron's Record Book shows how it was done.

'Tuesday 8th May. A new summer routine is being worked this term. Night flying cannot begin till 2130 of course, so students flying begins at 1315 and extends to the completion of night flying without a break. The only flying in the forenoon is devoted to the Formation Aeros practices. They launch at 1100 generally, so that the students have lectures, A.I. Trainer sessions, and Simulator, plus games or P.T. in the morning. Lt. McWilliam burst the starboard mainwheel tyre during the Stream Take-off for the formation aeros this morning, but made a successful landing and kept the Vixen on the runway, so it was soon back in the line of 'S' aircraft.'

52. With air brakes fully extended, a Vulcan B1 comes in to land. The Vulcans visited Yeovilton frequently during the 1960s. This was a security ploy when the operational Vulcans rarely spent more than one night on any particular air station. Their arrival was never 'officially' announced. Just 'popping in out of the blue', as the saying goes.

meaning, especially when the RAF cooks used to phone advanced orders to the local butcher, and he used to tip off his regular customer, Yeovilton.

Secrecy was also compromised during the Admiral's Inspection of Yeovilton in 1963. An Admiral's Inspection lasts for 24 hours, and the establishment involved has to suffer all manner of tests and trials of its capabilities and efficiency. The briefing of the Inspecting Officers of 1963 took place in Yeovilton's wardroom a few days beforehand. Unbeknown to those present the room had been bugged by enterprising members of the Electrical Department. A tape recorder in the ''Ladies''' next door took down every detail, and Captain Carver and his team were well briefed to repel intruders. Yeovilton's company of Ferret scout cars, borrowed from an army friend, wiped out the ''Admiral's Assault Party'' before it could attack the station. Rodney Carver only let the cat out of the bag in his retirement speech a year later.

When the Royal Air Force was formed in April 1918, it took over control of all forms of military aviation. After two decades of political wrangling the Admiralty regained control of naval aviation in May 1939. To celebrate the Fleet Air Arm's Silver Jubilee in 1964 a Royal Review was organised, at which HRH Prince Philip was to take the salute, on 29 May. Final rehearsals began a week earlier, and Yeovilton hosted 19 different squadrons and 150 aircraft of all types, from all stations of the Fleet Air Arm. The station's support facilities were stretched to the limit, accommodating and feeding all the visiting aircrew and ground crews, and providing all the essentials such as fuel, oil, oxygen and hydraulic fluid necessary to keep all those aircraft serviceable for the big day.

Unfortunately the weather turned foul, and the high performance aircraft such as the Buccaneer and Sea Vixens could not give their displays. The fly past was led by the historic Swordfish, piloted by Admiral Gick (who had flown one hundreds of times, including the

BISMARK attack). In the observer's position was Admiral Janvrin, who had navigated a Swordfish in the Battle of Taranto. He stood to salute as did Lt.Cdr. Topliss, who occupied the Telegraphist Air Gunner's place, where he had spent much of his war. Only the helicopters from Portland and Culdrose were able to follow with their display routines, but the day's events were nevertheless considered a great success, and all concerned well deserved the extra day's leave granted to mark the occasion.

The Royal Review was a proud day in the history of the Fleet Air Arm, and a special honour for Yeovilton in particular. The service had an enormous pride in its history, and in its present. The current mix of fighters (Sea Vixen and Scimitar), strike aircraft (Buccaneer), and anti-submarine and commando helicopters (Whirlwind, Wasp and Wessex) gave it enormous flexibility to operate from the carrier fleet anywhere in the world. An improved version of the Sea Vixen was just coming into service, and in June 1964 899 Squadron recommissioned at Yeovilton as the Sea Vixen FAW2 IFTU. Everything in the garden looked rosy.

53. Sea Vixen FAW2 of 899 Squadron, which reformed as the FAW2 IFTU, with the augmented selection of equipment and weapons which could be carried by this later version of Sea Vixen.

54. Two Sea Vixen FAW2s of 892 Squadron take off from Yeovilton in the late 1960s, each aircraft carrying only drop tanks on the outer underwing pylons.

55. The Sea Vixen's last fling. Six FAW2s of 892 Squadron formed the highly successful aerobatic team 'Simon's Circus', in 1968. Led by their CO, Lt. Cdr. Simon Idiens, and after only five weeks of practice, they put on an immaculate display of formation flying at the Biggin Hill Air Fair of 1968. By the end of the season they had flown forty aerobatic displays, and were regarded as one of the finest aerobatic teams of all time.

PHANTOMS

The General Election of 1964 returned a Labour government. A major Defence Review was undertaken by the new administration resulting in a general contraction of all the armed forces. The detailed discussions about the Royal Navy and Fleet Air Arm continued for many months. A ministerial pronouncement that the Royal Navy would not be committed to operations more than 150 miles from land undermined any argument to retain organic air cover for the Fleet, and sounded the death knell of the conventional aircraft carrier equipped with fixed-wing aeroplanes. That notwithstanding, the problems over Rhodesia's Unilateral Declaration of Independence sent the carriers to mount the ''Beira Patrol'', where Sea Vixens from Yeovilton's squadrons watched for tankers trying to supply contraband oil to Rhodesia.

The term ''aircraft carrier'' became political anathema to be replaced with ''through-deck cruiser'', a ship capable of operating helicopters in the anti-submarine and anti-ship roles. Far sighted optimists hoped that, in time, a naval version of the Harrier (at that time just entering service with the RAF), might be developed to operate from the small cruiser decks. Yeovilton itself barely escaped the axe.

The Royal Navy formed ''The Way Ahead'' committee. A Fleet Air Arm working party was one of the committee's subordinate bodies, whose job was to assess future naval aviation plans and resources. It proved extremely difficult to demonstrate to the Secretary of State, Mr Dennis Healey, that three active air stations were required to support an all-helicopter Fleet Air Arm. The bases at Culdrose and Portland were adequate to support ASW helicopter squadrons, but a third was needed to house the commando helicopter squadrons. The choice fell between Brawdy (in Pembrokeshire) and Yeovilton.

After a detailed assessment, Yeovilton was selected because of the proximity to Salisbury Plain with its associated ranges, and better geographical communications. An added advantage was that Yeovilton could also handle the fixed-wing aircraft of the Fleet Requirements Unit, which, if moved from Hurn Airport near Bournemouth, would still be within range of the Fleet exercise areas off Portland.

This final plan was initially rejected by the Ministry of Defence, and Yeovilton's future continued to be in doubt for some time. Many months later the navy's argument was finally accepted, and Yeovilton was saved. There followed a difficult period for the station personnel in simultaneously supporting the last fixed wing units of the Fleet Air Arm, while preparing the station to receive the Commando Helicopter Squadrons. Those last fixed-wing elements included the ultimate Fleet Air Arm jet fighter, the McDonnell Douglas Phantom F4K, and Yeovilton underwent many changes in the late 1960s because of that powerful and complex aeroplane.

The development of the station to receive Phantoms began long before the first one arrived, and continued alongside the Sea Vixen squadrons going about their normal duties. The front-line task of the Vixen was coming to an end, but 892 Squadron decided that their demise should be with a bang, not a whimper. Going one better than "Fred's Five", six Sea Vixen FAW2s of 892 Squadron made up "Simon's Circus", led by their CO, Lt.Cdr. Simon Idiens. All but four of the aircrew were on their first front line tour. Their average age was only 26, and the youngest member was a 21 year old pilot.

After only five weeks practice they made their first public appearance at the Biggin Hill Air Fair of 1968, and put on an immaculate display of formation flying. By the end of the season they had flown forty aerobatic displays, and become known as one of the finest aerobatic teams of all time.

Simon's Circus also made local headlines in other ways. "Sailors swab decks in navy blue towns" headed one newspaper story. During practice around Yeovilton an aircraft's smoke canister malfunctioned and the prevailing winds spread blue dye over a wide area. Car paintwork and laundry lines took on a blue polka dot aspect, but clean up squads were sent out to put things right, and an official apology was accepted by everyone affected. The team adopted a lion cub from Longleat as a mascot, and often had their photographs taken with it. As the cub grew a replacement was acquired, but this second one had a distinctly awkward temperament, and regularly tried to bite lumps out of the airman holding it.

The workmen building the new Phantom facilities often had Simon's Circus as a diversion from their long list of projects. A new mains electricity supply had to be installed. Every workshop was upgraded: the Engine Repair Section became the major centre for Rolls Royce Spey engines, supported by an engine test bay capable of handling the Spey's awesome full-throttle power; the Electrical and Radar Workshops were enlarged to provide a new laboratory for the Phantom radar and weapons

56. A newly arrived Phantom about to enter the Spey Engine Test Bay, specially constructed for the Phantom and its awesome noise and power. The test bay is essentially a giant silencer.

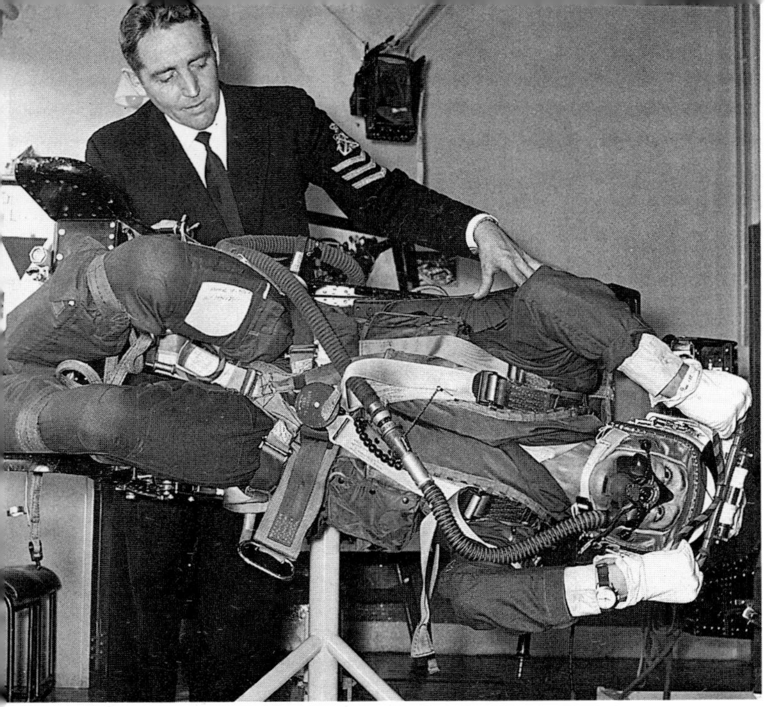

57. Ejector release practice. Lt. Nick Child suffers on the Phantom Ejection Seat Training Rig, at the hands of a Safety Equipment Petty Officer.

systems, such as the Air Data Computer, Flight Control and Bombing Computer and Sparrow missiles; a combined Safety Equipment and Ordnance facility was created to deal with ejection seats, parachutes and dinghies; the capacity of the bulk fuel store was increased by 180,000 gallons, with improved facilities for bowser control and maintenance in a new combined centre.

McDonnell Douglas provided the initial spares support. They were given 5,000 sq. ft. of storage space in the Phantom hangar, with offices in an adjacent wooden hut, and Westinghouse required a temperature controlled store for their Radar Support Unit. Living space had to be found for around 400 extra personnel. Two new ratings' mess blocks were built, and even then 160 temporary beds had to be provided in caravans attached to existing buildings.

More buildings were required to house the two simulators which formed an essential part of Phantom

58. Pre-flight checks prior to a sortie aboard XT858/724-VL of 700P Squadron, the Phantom IFTU. External access ladders are fitted, although the crew could enter the cockpit via the internal steps, of which the hatch cover can be seen just behind the foot of the external ladder. The fuselage is adorned from end to end with non-standard stencilling to aid the training of ground crews and engineering staff. The air inlet duct behind the radome provides cooling for the complex avionics.

aircrew training. The Observers' or RIO (Radar Interception Officer) Simulator, made by Redifon UK, was commissioned in late 1967, and the Aircraft Simulator, bought from Goodyear in the USA, followed about a year later. Both were superb pieces of equipment which made major contributions to the Phantom programme.

An exemplary Technical Training School was opened in February 1968. Housed in a hangar dedicated almost entirely to the Phantom, it had six laboratories on the ground floor and nine classrooms above. The technical instructors went on courses with all the main Phantom contractors and also with the US Navy, for operational experience. Besides turning their specialised knowledge into lecture notes, they also built a multiplicity of training aids, which demonstrated the complex workings of the Phantom's mechanisms and systems to the groundcrew trainees.

The airfield itself, which had been fitted with arrester gear for emergencies for many years, was closed for two months while an upgraded system, PUAG, (Purpose Use Arrester Gear) was installed at each end of both runways.

The Phantom IFTU, 700P Squadron, formed in April 1968 under Cdr. A.M.G. Pearson. During the high speed trials, Lt.Cdr. Bob Evans, the Maintenance Test Pilot, achieved Mach 2.1, more than twice the speed of sound, without any damage to the aircraft.

59. Phantom XT862/722-VL of 700P Squadron, in its natural environment - 'going at it' over the sea, during flight trials early in 1969.

60. Yeovilton groundcrew innocently servicing a new RAF Phantom on its way from the USA to RAF Aldergrove, having already 'zapped' the fin with 'Fly Navy' in six inch high dayglo letters.

The IFTU serviced two RAF Phantoms which arrived at Yeovilton from the USA in formation with two naval Phantoms. The ferry crews did not notice that their aircraft had been "zapped" during the turnaround at Yeovilton, but RAF Aldergrove were not amused to find their brand-new aeroplanes carrying six-inch high dayglo stickers on the fin imploring "Fly Navy".

In January 1969, 767 Squadron recommissioned at Yeovilton as the Phantom Conversion Unit. The squadron first trained its own instructors, and then they trained aircrew both for 892 Squadron (which formed on Phantoms around a nucleus of 700P Squadron personnel at Yeovilton in March 1969), and for the RAF's Phantom fleet. A course usually consisted of two RN and two RAF crews, who each spent about 70 hours flying the aircraft. The unit disbanded in August 1972 having trained 70 RN, 30 RAF, 2 US Navy and 2 US Air Force officers to fly and fight the Phantom. The preparation put into technical training had paid off. The maintainers had achieved a superb serviceability rate, with their eleven Phantoms flying over 6,000 hours in just over three years. The squadron commander, Lt.Cdr. Peter Marshall had been awarded the Air Force Cross for his skill and gallantry in bringing back a Phantom without a nose cone, which had broken away in the air, and the squadron itself had been awarded the Boyd Trophy for its outstanding work with the Phantom.

61. The end of an era. 892 Squadron's Phantoms leave Yeovilton in July 1972, bound for their new home base at RAF Leuchars.

62. XT865/011 of 892 Squadron is catapaulted from ARK ROYAL, the only Royal Navy carrier able to operate the Phantom.

892, the last conventional fixed wing fighter squadron in the Fleet Air Arm, was parented by Yeovilton until July 1972, when it moved to RAF Leuchars.

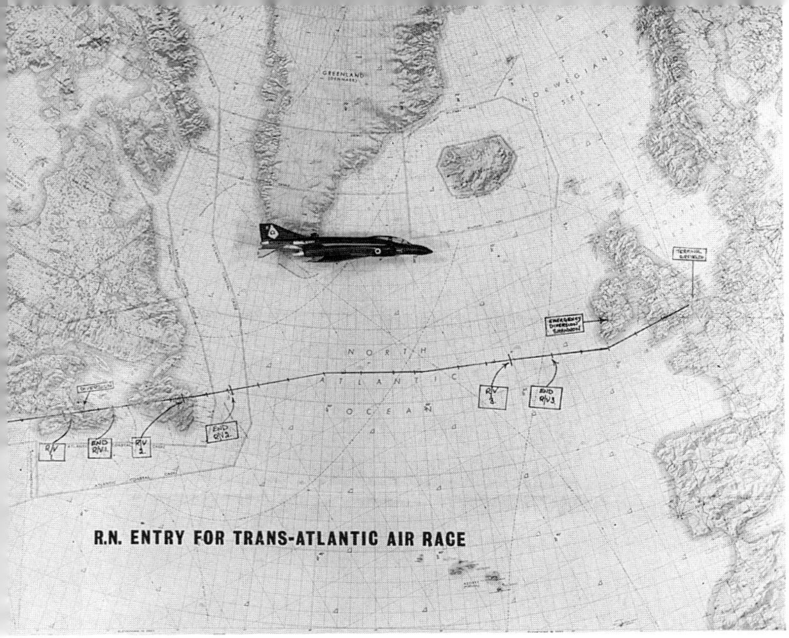

NEW YORK / LONDON

63. Schematic map of the route used by 892 Squadron's Phantoms in the 1969 Air Race. In-flight refuelling positions are marked as R/V (Rendezvous).

64. Lt. Cdr. Peter Goddard, Senior Observer, (left), and Lt. Cdr. Brian Davies, CO of 892 Squadron, checking over the final details before making their record breaking and race winning run across the Atlantic in XT859.

TRANSATLANTIC AIR RACE

In 1919 John Alcock and Arthur Whitten Brown had flown a Vickers Vimy bomber from Newfoundland to Ireland in the first non-stop air crossing of the Atlantic Ocean. To mark the fiftieth anniversary of their achievement the Daily Mail, which had established a prize for that first crossing, instituted a transatlantic race for the fastest time between the top of the Empire State Building in New York and the top of the Post Office Tower in London. The official British Services entry centred around the Phantoms of 892 Squadron from Yeovilton.

In overall charge of the attempt was Captain Raymond Lygo RN, later Admiral Sir Raymond Lygo and Chief Executive of British Aerospace. At that time Captain Lygo was serving at RAF Strike Command HQ, and he devised a scheme involving an RAF Wessex helicopter in New York, three Phantoms, the Handley Page Victor tankers of 55 Squadron from RAF Marham and a Wessex helicopter from 707 Squadron, RNAS Culdrose. The Phantom observer rode pillion on a BSA motorcycle through New York to the 30th Street West heliport, there to embark in the Wessex which flew to Floyd Bennett Field where the Phantom was waiting. Immediately after

65. The winning aircraft, XT859/001-R, taking fuel from Victor K.1A, XH648, of 55 Squadron from RAF Marham.

66. Lt. Hugh Drake, observer of Phantom XT861/003, sprints to the waiting Wessex of 707 Squadron from RNAS Culdrose, for the Wisley to London leg on May 7th 1969. The Phantom pilot on this second run was Lt. Alan Hickling.

take-off the Phantom climbed at full power to 33,000 feet, ''cruised'' to Nantucket Island at 650 mph, before accelerating to 1,100 mph at 45,000 feet towards the first refuelling rendezvous with a Victor tanker 60 miles south of Nova Scotia. Having slowed down for the tanker hook up, the Phantom accelerated for a second supersonic leg to Newfoundland. A second tanker rendezvous was made to top up the Phantom's tanks, before a long transatlantic cruise at 650 mph at 39,000 feet, along a great circle route.

West of Ireland the Tribal class frigate, HMS NUBIAN, acted as ''air traffic controller''. International air traffic rules demanded that the Phantom's descent to 33,000 feet and third Victor hook up was done under the strictest control, in the middle of the heavily used commercial air lanes. The Phantom then made a final supersonic dash round the south of Eire, decelerating to subsonic speed 45 miles west of Lundy Island, and running on to BAC's airfield at Wisley at 700 mph. The observer was out of the Phantom almost before it had finished its landing run, transferring to the waiting Wessex that whipped him away to the Moss Construction site immediately adjacent to the Post Office Tower. When the Wessex landed he had an 85 yard sprint to the

67. Captain Raymond Lygo poses with members of the Trans Atlantic Air Race Team. Aircrews from the left are Lt. Cdr. Doug Borrowman and Lt. Paul Waterhouse (Run 1), Lt. Cdr. Peter Goddard and Lt. Cdr. Brian Davies (Run 3, the winners), Lt. Alan Hickling and Lt. Hugh Drake (Run 2). At the time of the photograph the aircraft were still in 700P Squadron markings, 722/ XT862 and 727/XT861.

68. The Wessex V alights on the temporary landing pad, while the observer makes ready for his dash to the lift.

lift which took him to the finishing point at the top of the Tower.

Three Phantoms positioned to Floyd Bennett Field on 24 April 1969, and joined up with their maintenance team. To set a target time and examine the workings of the plans in practice, the first attempt was made on the first day of the race, 4 May. Lt. Paul Waterhouse, piloted by Lt.Cdr. Doug Borrowman, achieved an overall time of 5 hours 30 minutes 24 seconds. Three days later on the 7th, Lt. Alan Hickling flew Lt. Hugh Drake to a time of 5.19.17. On the last day of the race the ''bosses'', Senior Observer Lt.Cdr. Peter Goddard, and Commanding Officer Lt. Cdr. Brian Davies, put the issue beyond doubt, winning the race with the fastest overall time of 5.11.22. En route they had set a new record across the Atlantic of 4.46.57. It was, however, a team effort, and a remarkable performance by all the elements involved in a complex but exhilarating undertaking.

As a peacetime inter service deployment and training exercise the Transatlantic Race had the benefit of a clearly defined and high profile objective. The win was the result of sound planning and accurate implementation, and the subsequent publicity for the then unfashionable Services was an additional reward.

HELICOPTERS AND HARRIERS

Despite the departure of the front-line jet squadrons, fixed-wing aircraft were still very much part of Yeovilton's activities in the early 1970s. The Fleet Requirements Unit moved up from Hurn in the autumn of 1972, and Station Flight had its Sea Herons and Devons. But these were only second line units, and Yeovilton's overriding responsibility was to prepare the station to accept the Commando Helicopter Squadrons.

A very close liaison was established with RNAS Culdrose to ensure that the squadrons would transfer into a well prepared new home. Additional hardstandings and new office blocks were built, and Air Traffic Control had to develop new circuit patterns and control procedures to handle safely the mix of rotary wing and fast jet traffic. The windsocks were repositioned, taxying routes were repainted, and the Visual Control Position atop the control tower was modified to allow better surveillance of the complex airfield situation. Minor changes were made to the Radar Room, so that controllers could keep track of helicopters within the county rather than supersonic jets a hundred miles away. In 1974, further modifications were made to the radar to improve the height indication, and closed circuit television was instal-

69. De Havilland Sea Heron C1, of Heron Flight, formerly Station Flight, overflying Chesil Beach in 1986. Sea Herons saw continuous service with the FAA as VIP transport aircraft from 1972 to 1989.

70. The Control Tower's Visual Control Position in March 1976.

The Duty Air Traffic Controller is at his desk, looking at the approach of runway 09 through his binoculars. The Local Airfield Controllers are monitoring radio transmissions.

71. In May 1972 members of 707 Squadron, encumbered with suitcases and baggage, leave their Wessex HU5s on the hardstanding at Yeovilton after flying in from Culdrose. First job next day is to change the CU tail markings to VL.

led as part of an upgraded aircraft movements and weather information system.

In May 1972, 707 and 846 Squadrons, equipped with Westland Wessex HU5 helicopters, arrived at Yeovilton. Commando helicopter pilots all went through 707 Squadron's Operational Training Course, and 846 Squadron was the Headquarters and Trials Squadron for the Wessex 5. By the end of the year the two Wessex front line squadrons, 845 and 848, had also transferred to Yeovilton. The airfield at Merryfield, near Taunton, was reactivated to allow the specialist commando operations to be practiced without further complicating the airfield movements at Yeovilton.

During their 20 week course in 707 Squadron, RN and Royal Marine pilots were taught the special skills required for ''brush fire'' operations, in which troops had to be delivered to or collected from all kinds of terrain. They learned mountain flying in Wales, or from a US base in the German Alps at Bad Toelz. A regular deployment was to Detmold in West Germany for familiarisation exercises in the mid-European theatre of war. There was also deck landing training in the Royal Fleet Auxiliary vessel, ENGADINE. The course ended with a week's military exercise on Dartmoor, living under canvas and practicing with the commando troops.

In September 1974 the squadron was honoured to

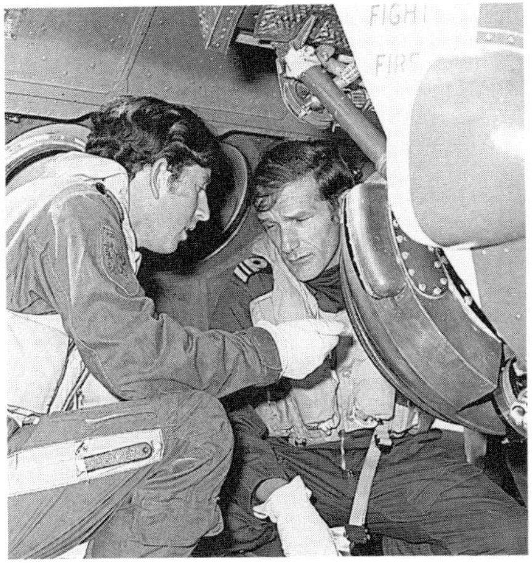

72. Getting down to basics in the nose of a Wessex.
Lt. Cdr Alan McGregor undergoing some intense questioning.

73. Lt. HRH the Prince of Wales talks to Cdr. C.R.V. Doe, Cdr.(Air) of RNAS Yeovilton, and Lt.Cdr. P. Voute, CO of 707 Squadron, on the occasion of the Squadron's tenth anniversary ceremonial flypast in December 1974. Prince Charles led the flypast, having just completed his three months training course as a Wessex pilot.

have a special course member. "Red Dragon Flight" of two aircraft was formed within 707 Squadron to train HRH The Prince of Wales, who at that time was serving as a naval lieutenant. The Prince completed his course successfully in December 1974, which coincided with the squadron's tenth anniversary in its present role. All but one of its past commanding officers was able to attend a special celebration, and to take part in a flypast of sixteen Wessex in diamond formation, led by Prince Charles. The two Red Dragon helicopters streamed the Royal Standard and the White Ensign. By the time he left to join 845 Squadron, Prince Charles had earned the Double Diamond Trophy, presented at the end of every course to the pilot making the most progress.

The front-line commando squadrons continued to exercise in Britain and Europe, in addition to providing squadrons aboard the commando carriers, BULWARK and ALBION, and flights aboard RFA vessels, such as REGENT, RESOURCE, TIDESPRING and OLNA. One of the most demanding exercises was, and still is, "Clockwork", based in Northern Norway. Here aircrew are trained to fly, and groundcrew to maintain, helicopters in temperatures as low as -40C. The base camp is at Bardufoss, a Forward Air Base, but part of the training takes place when an independent unit operates from a Forward Operating Base (FAB), where survival training is critical for such extreme conditions.

An important new helicopter joined the Fleet Air Arm in the 1970s, the Westland Lynx, destined to replace the Westland Wasp in the small ships flights. The Lynx IFTU (700L Squadron) was formed at Yeovilton in September 1976, a joint venture with the Royal Netherlands Navy. Their task was to establish the aircraft's limits and capabilities in all roles and all weather conditions. In just over thirteen months the unit achieved 4,250 flying hours on eight aircraft without a fatal accident.

The Lynx incorporated a great deal of new technology, giving a dramatic improvement in capability over the Wasp. The Rolls Royce Gem engines were brand new, as was the Sea Spray nose radar and Tactical Air Navigation System. All of these had to be put to the test. The Dutch contingent, which had been happily integrated into the unit, left in the summer of 1977, and 700L Squadron added the development of the Sea Skua anti-ship missile system to its tasks. The Squadron received the Boyd Trophy for its work before being absorbed into 702 Squadron, which reformed in 1978 as the Lynx HQ and Training Squadron, responsible for aircrew conversion on to the type and for further operational development of the helicopter.

74. The unofficial badge adopted by 700L Squadron, the Anglo-Dutch IFTU for the Westland Lynx.

75. In 1976 a brand new helicopter arrived from the factory down the road. The Westland Lynx was to replace the Wasp in the small ships flights, after undergoing trials at Yeovilton with 700L Squadron.

700L Squadron, the Lynx IFTU, at work and play.

76. Finding out what can be done with a Lynx, like stowing it aboard a Hercules of RAF Support Command,

78. Finding out about milk from the National Dairy Queen comes under the heading of 'play'.

77. lifting it up by its rotor, after a heavy emergency landing, due to hydraulic failure, had deposited it firmly in the mud.

Ship's Flights were formed from 702 Squadron, and parented by the unit until the task was taken over by 815 Squadron in January 1981.

By 1979 the Wessex 5 had been in service for nearly fifteen years, and new equipment for the commando squadrons was in the pipeline. During the year three pilots from 846 Squadron learned to fly the Westland Sea King HC4 at Culdrose, and thus the squadron was prepared when the first of its new helicopters arrived in the autumn of 1979. For their first "Clockwork" at Bardufoss, four Sea Kings flew over via Lossiemouth and Bergen to join the Wessex already in Norway. The Sea King's initiation into arctic conditions ended with exercise "Anorak Express", during which both Wessex and Sea Kings operated together from a Forward Operating Base (FOB) for the first time. This worked out well and set the pattern of operation for succeeding years.

The Sea Kings changed fundamentally the operating procedures of the commando squadrons. They could now fly an aircraft with ten maintainers

83. HERMES, flagship of the Falklands Task Force, getting under way with Yeovilton's 800 Squadron (Sea Harriers), 845 and 846 Squadrons (Sea Kings). The leading Harrier on the ski-ramp is one of those retrieved from RAE Bedford, where it had been undergoing trials.

such as the Lynx from 702 Squadron and the Sea Herons and Sea Devons of Heron Flight flew all over the country with stores and important passengers. The station set up a Welfare Information Cell, manned around the clock, to answer the problems and anxieties of families whose husbands and fathers had been rushed away to the South Atlantic.

The Falklands could not have been retaken without naval airpower. The commando helicopters in particular performed prodigious feats of troop insertion and supply, often under fire or in dreadful weather conditions. The Americans had developed the idea of "airborne cavalry" in Vietnam, and the Falklands was the first time that British forces had tested the approach in battle conditions. Yeovilton's two commando squadrons, 845 and 846, had aircraft in 18 different ships of the Task Force, as well as detachments ashore. A third "emergency" unit, 848 Squadron, was formed at Yeovilton from 707 and 772 Squadrons, but its battlefield effectiveness was diminished by the loss of several aircraft aboard ATLANTIC CONVEYOR.

The essential nature of helicopter support is demonstrated by the statistics of 846 Squadron. In the three months April - June 1982, their eleven aircraft flew for 3,107 hours in 1,818 sorties of all kinds. On the first day of the re-invasion seven helicopters lifted nearly one million pounds of stores and 520 troops ashore. Despite the bad weather, and crowded conditions aboard their home ships, the squadron's maintainers achieved an 85% record of serviceability overall, and for the final advance on Port Stanley, 100% of 846 Squadron's aircraft were fit to fly. The groundcrews' performance is even more remarkable when one considers that the flying hours in those three months were more than the squadron would normally clock up in a whole year.

84. '....and we'll all come up from Somerset'

Sea Kings over Falkland Sound, May, 1982.

Painted by David Gibbings as a tribute to the part played by Yeovil in the Falklands Crisis.

The Sea Harriers were also worked hard. The two front line squadrons absorbed additional aircraft and pilots from 899 Squadron, and embarked in the carriers HERMES (800 Squadron) and INVINCIBLE (801 Squadron). Once they were inside the Total Exclusion Zone around the Falklands, the two squadrons maintained a permanent Combat Air Patrol (CAP) around the Task Force, as well as mounting air strikes against Argentine positions, both prior to and during the re-invasion. They were reinforced in early May by 809 Squadron, the emergency unit formed on 8 April 1982 by scouring the country for experienced Sea Harrier pilots and serviceable aircraft. By dint of incredible work from many quarters and hard training at Yeovilton, the squadron flew out only 22 days after its formation taking another eight Sea Harriers to Ascension Island. From there they shipped aboard ATLANTIC CONVEYOR, and fortunately had flown off to the other carriers before the freighter was tragically sunk by an Argentinian Exocet missile.

The Sea Harrier squadrons scored over twenty air combat victories during the war, with no combat losses to themselves. At the time the media suggested that the

85. Arming a Sea Harrier with the Sidewinder AIM-9L air-to-air missile, aboard HERMES. This latest version of Sidewinder was one of the four factors accounting for the Harrier's success during the Falklands War, the other three all being human.

manoeuvre called ''viffing'' had been a major factor in this success. Viffing is an abbreviation for ''Vectoring in Forward Flight'', in which the Harrier pilot, when flying normally, turns the jet nozzles downwards to put the aeroplane into a very rapid and almost vertical climb, designed to put off the aim of an attacker. Nothing could be further from the truth, even if the truth is too mundane for some of the tabloid press.

There were four major elements in the Sea Harrier's success. One was the use of the most modern version of the Sidewinder missile, the AIM 9L, but no weapon is any good on its own without a competent system of control and delivery. This was provided by the other three factors, all of which were human. The basic rules of air combat were laid down in the First World War. A fighter pilot's best friends are height and speed. Von Richthofen preached the maxim, ''Always above, seldom on the same level, never beneath.'' The Sea Harriers on CAP were handled by Direction Officers, based in ships of the Task Force, who had been trained in Yeovilton's Direction School. The Directors used their radar information to place the Sea Harriers in the best positions to intercept incoming raiders. The rest was down to the pilots.

Most of the Argentinian pilots were fairly young men, but they were pitted against a high proportion of very experienced pilots in the Sea Harriers, many of whom had flown Phantoms, and some had experience stretching back even to the Sea Vixen. They had thousands of hours in fast jets and were supremely confident in their flying abilities. Their skill and leadership rubbed off on the younger Sea Harrier pilots, and all of them had the advantage of the fourth critical factor, the most modern tactical methods during their peacetime training. An American aviation journalist visited Argentina after the Falklands War, and later reported his findings to a gathering of Yeovilton's officers. He was quite specific in his statement, ''The Argentinian pilots are good 'stick and rudder' men, but they have very few tactical skills''.

Yeovilton provided 126 aircraft, 200 officers and 1,200 ratings for the Task Force. Altogether, sixty one decorations and honours were awarded to Yeovilton personnel.

Sadly, a number of Sea Harrier and helicopter aircrew were killed during the Falklands War. Those lost from Yeovilton are commemorated at their home station by twelve beech trees, planted on the sports field, one for each man who did not come home with the victorious squadrons.

86. Mrs Betty Williams, wife of the Captain of HMS HERON, unveiling a plaque commemorating the twelve members of the Ship's Company who were killed during the Falklands War.

POST FALKLANDS

The demands of the Falklands War had demolished the normal routine at Yeovilton. Those aircraft, instructors and groundcrew left behind had done their best to maintain the training programmes, but it had been difficult. When the squadrons returned most of the aircraft were well overdue for major overhauls, and it was some months before the rhythms returned.

By the summer of 1983 the Royal Navy was able to participate in the major NATO exercise ''Ocean Safari'', and ILLUSTRIOUS and HERMES were deployed to the eastern Atlantic. On the afternoon of 6 June S/Lt. Ian Watson of 801 Squadron took off in his Sea Harrier from ILLUSTRIOUS, operating off the Iberian Peninsula. After carrying out his planned sortie he was returning to the ship when he suffered a total failure of his navigational aids and radio. Unable to locate the carrier and with his fuel running low, Watson used his radar to search the sea below. Only one small ''blip'' showed up, and the pilot headed straight for it. The ship was a Spanish freighter, ALRAIGO, whose crew were astonished when the Sea Harrier appeared from nowhere and carried out a vertical landing atop the containers loaded just forward of ALRAIGO's bridge.

As the Sea Harrier settled vertically it began to roll backwards and Watson retracted the undercarriage to halt it. The rear fuselage struck the roof of a Mercedes van parked on deck, and one of the van doors flew open, supporting the tail of the aeroplane. The Spanish crew helped Watson out of his aeroplane, and lashed it securely to its unusual landing pad. The captain radioed his base about the new addition to his deck cargo, and ALRAIGO continued on towards its destination of Tenerife.

When the news was received at Yeovilton, Cdr. Paul Madge was instructed to take an engineering team under Lt. G.J. Wilson to Tenerife to recover the Sea Harrier from ALRAIGO. When they arrived at their hotel in the port of Santa Cruz, they were astonished to find a waiting horde of media men from all over the world. This was only one of the problems the team had to face.

The ALRAIGO docked on 9 June, and from the dockside a preliminary assessment suggested it would take two or three hours to remove the Sea Harrier from the deck. S/Lt. Watson was relieved to see the naval party because, although he had been very well looked after, he could not speak Spanish and not one of the 14-man crew could speak a word of English. Watson was sent back to ILLUSTRIOUS via Yeovilton, and apart from being dubbed ''The Alraigo Kid'' and ''Spanish Fly'', he was the lucky one!

The next few days verged on farce. The Spanish crew were concerned that their shipowner would be the only beneficiary of the salvage money due on the £7 million Sea Harrier. Although the Captain might let Wilson's team aboard to work for a while, someone from another faction would then force him to order them off. All the while Cdr. Madge and British Consular officials were negotiating with a bewildering variety of vested interests to secure the release of the Sea Harrier. They would gain a concession, Wilson's team would board ALRAIGO, do a little more work, be ordered off and negotiations would begin again.

Finally the naval team were allowed to finish the job, a mobile crane arrived, and at long last the Sea Harrier was lifted from the ALRAIGO and onto the dockside. The two hour job had taken eight DAYS. Even then, the local dockworkers shop steward warned Madge that if his team took much longer it would be siesta time when all work must stop. But the Sea Harrier was towed around the quay and loaded onto the BP tanker BRITISH TAY, which took it back to Britain and repair at Yeovilton. Greatly relieved, Madge sent his final signal to the Ministry of Defence, ''Grey canary flies BP - OK. Signed Senior Naval Officer Tenerife Atlantic Island Region (SNOTAIR).''

The summer of 1983 in Somerset was almost as hot as Tenerife, and beautiful weather accompanied an exchange of helicopter squadrons. Yeovilton's Lynx squadrons,

87. "Thanks for the lift." S/Lt Ian Watson takes his leave of ALRAIGO's captain.

88. Organised chaos, and everyone wants a piece of the action.

89. On the move, at last.

90. 001 pulling crowds on the quayside at Tenerife, before starting its journey home to Yeovilton.

702 and 815, transferred to Portland in July, making room for the 3 Commando Brigade Air Squadron (3 CBAS) of the Royal Marines, which moved in with their Gazelle and Lynx helicopters from their former base at Coypool. There was also another organisation new to Yeovilton, the Indian Navy Training Unit, INTU.

In 1981 the Indian Navy had announced its intention to purchase six Sea Harrier single seater and two two-seat training aircraft. It was agreed that the Royal Navy would train their pilots and maintainers at Yeovilton, and INTU was formed in October 1982. Fourteen Royal Navy officers and senior ratings trained 164 Indian Navy personnel to fly and maintain their own Sea Harriers. One Indian air traffic control officer was attached to Yeovilton ATC, and he learned to talk down his pilots in both Hindi and Telagu, the dialect of his home state of Andhre Pradesh.

By the end of 1983 the Indians had become a completely self-contained squadron with six Sea Harrier FRS51 and two Sea Harrier T60 aircraft. Just before Christmas leave began, 300 Indian Navy Squadron, the White Tigers, under Cdr. Arun Prakash, ferried their aircraft to Goa. Prakash soon made the first deck landing aboard INS VIKRANT on 20 December 1983.

Although the two very different cultures had worked happily together, there had been minor problems when "Jack" had been unable to wrap his tongue around some of the more complicated Indian names. In a mutually acceptable compromise, Petty Officers Subramanian and Debramanian were known universally as "Supper Money" and "Dinner Money".

91. The Indian Navy taking final delivery of six Sea Harrier FRS51s and two T60 trainers, their pilots and maintainers having completed a rigorous training course at Yeovilton. On the right of the picture Vice Admiral Sir John Cox, FONAC (1982/3), waits expectantly for a piece of cake.

92. Three Sea Harrier FRS1s of 899 Squadron (nearest camera), with four belonging to INTU, ready for take off on a training sortie. The fire extinguishers on the two-wheeled trolleys are part of groundcrew's standard equipment.

WORKING ON

There was no Air Day in 1983, not because of the Falklands aftermath, but because the continuous programme of facility improvement required that new runway lighting be installed. All conventional fixed wing flying moved away from Yeovilton for the summer months, but the Sea Harriers and helicopters continued to operate from the taxiways and grass areas. For the benefit of station personnel, additional squash and tennis courts were built and a start was made on new Wrens' quarters.

At the end of 1983 Yeovilton became the host station of a NATO unit, the Maritime Electronic Warfare Support Group. MEWSG has no aircraft, just countless mysterious boxes, whose job is to simulate the communications jamming that might be experienced in a war scenario.

Until the mid 1980s Yeovilton's water system had owed a lot to 1940's plumbing, and there were areas of the station where the water pressure was so low that a fire hose would barely fill a kettle. So a new system was installed with the mains buried underground, but the water tower was plain for all to see, variously christened "The Tipsy Wine Glass" or "The Admiral's Goblet".

The water tower replaced an old landmark, the Type 965 radar scanner, adorned "Fly Navy", which stood alongside the control tower for many years. In 1985/6 the radar system was modernised, and a new Precision Approach Radar was installed. Fitted on an automatic turntable, it allows the controllers to alter the radar cover of the duty runway in 30 seconds, compared with the old system's 15 minutes. The control tower was itself refurbished, with a new Radar Approach Room and the latest display and control equipment. Work began on reshaping the old V-bomber South Dispersal area, to allow for the building of two new hangars to house FRADU's aeroplanes, many of which had previously been parked outside.

The training tasks have continued right through the 1980s, with improvements being made as resources permitted. By 1985 all of 707 Squadron's training was being carried out on Sea King helicopters, and the Sea

93. New water tower under construction during 1984/5. FRADU's Canberras are parked nearby, three TT18 target towing aircraft on the left and a T22 threat simulator on the right.

94. Alterations to the control tower in 1985. The Type 965 radar scanner has been dismantled, and its 'FLY NAVY' has moved to the roof, but fashions change and it's not there now. The signal square on the ground by the car park used to show the active runway by the direction of the T, but is no longer used.

Harrier squadrons began to send detachments to the Air Warfare Training Ranges at Decimommanu in Sardinia, where sophisticated electronic surveillance systems can record every detail of mock air combat.

By 1986 845 Squadron was the last front line squadron using the venerable Wessex 5. The unit had received its first Sea King early in the year and had completely re-equipped by the end. In May two Sea Harriers of 800 Squadron flew unsupported via Iceland, Greenland, Canada and North Carolina to the US Naval Air Station at Pensacola, Florida (where many of Yeovilton's Second World War trainees had received their initial flying training) to participate in the 75th birthday celebrations of US Naval aviation. In return, Yeovilton hosted the Catalina flying boat, which followed the Atlantic route flown by four US Navy flying boats in 1919.

At the end of 1989 Yeovilton said goodbye to a number of aircraft which had almost become permanent fixtures - the Sea Devons and Sea Herons of Heron Flight. The original organisation had been known as Station Flight, and for many years this had operated a mixture of aircraft, including a Wessex SAR helicopter in the early 1970s. In 1972 the first Sea Heron arrived, to provide transport for FONAC and other VIPs in support of the smaller Sea Devons. When the Fleet Air Arm communications squadron, 781, was disbanded at Lee on Solent in 1982, its aircraft were transferred to Yeovilton. By this time the unit was known as Heron Flight and its responsibilities also included the servicing of all visiting aircraft, and the running of the Historic Flight. The Devons and Herons were not new when they joined the Fleet Air Arm and their end was inevitable. The familiar drone of their Gipsy Queen piston engines has been replaced by the turboprop whistle of two British Aerospace Jetstreams.

In 1990 Yeovilton is the Royal Navy's largest aviation establishment, employing over 3,000 people, service and civilian, including more than 200 Wrens. It is one of the busiest military airfields in the UK and one of the largest in Europe, with nine squadrons and over 100 aircraft. It is the only one to operate helicopters, fixed-wing and jet V/STOL aircraft simultaneously. That muddy lagoon has changed a lot since 1940.

95. Avro Vulcan B2 XL426 lands at Yeovilton to refuel in 1985.

96. The Fairey Swordfish LS326 of the Royal Navy Historic Flight, complete with torpedo. This aircraft had been maintained by Yeovilton's Station Flight for many years before the Historic Flight was founded in 1972.

97. The Fairey Firefly, WB271/204, which was located in Australia by officers of VICTORIOUS in 1966, and bought with a whip round in the wardroom.

98. The Fairey Firefly in formation with the first production Hawker Sea Fury FB11, TF356/123, which was rebuilt by Hawker Siddeley in 1972.

99. The two-seat Sea Fury T20, which joined the Historic Flight from Germany in 1977, in formation with the single seat version.

100. Hawker Sea Hawk, WV908/188, Historic Flight's only jet, which was transferred from RNAS Culdrose in 1982

ROYAL NAVY HISTORIC FLIGHT

The Royal Navy Historic Flight is a separate organisation from the Fleet Air Arm Museum. Its aircraft are still on the Navy's charge, and they are cared for and flown by members of the Fleet Air Arm.

Although the RN Historic Flight was formed as such only in 1972, for many years prior to that, Yeovilton's Station Flight had maintained an airworthy Swordfish, which was a popular visitor at air shows all over the country. In 1972 the Swordfish, having flown for 460 hours over eleven seasons, had a major rebuild at the Royal Aircraft Establishment, Farnborough. Fairey Engineering at Manchester's Ringway Airport found two brand new Swordfish propellers and donated them to Yeovilton. In the same year there was an auction at a Canadian farm, which had several Swordfish lying in its fields, where they had been since the farmer had bought a job lot of war surplus in 1946. One of the best was bought by the Canadian Air Force on Yeovilton's behalf, to provide a spare engine, and was transported to Britain by the RAF.

The Swordfish was thus assured of several more years life, and it was joined in 1972 by the first production Hawker Sea Fury FB11, which was rebuilt by Hawker Siddeley, supported by many other aerospace companies which had supplied systems for the original production run. These two aircraft formed the basis of the RN Historic Flight.

Whilst in Australia in 1966, some aircrew officers from VICTORIOUS had located a Fairey Firefly, and a whip round in the wardroom had enabled them to buy the aeroplane and present it to the Fleet Air Arm Museum. The Firefly was later rebuilt to airworthy condition by CPO Ron Gourlay and the staff of Historic Flight. March 1973 saw the 30th anniversary of the first flight of the first production Firefly from Fairey's Great West Airfield, which had now become Heathrow Airport. Yeovilton's Commanding Officer and Operations Officer marked the occasion by flying the Firefly to Heathrow, and afterwards the Firefly became a permanent member of the Historic Flight.

More aircraft were added over the next few years: a Tiger Moth in 1974; a two-seat Sea Fury T20 (from Germany) in 1977, for pilot training; and in 1982 the Flight's only jet, a Hawker Sea Hawk was transferred from RNAS Culdrose.

The maintainers' task was always difficult and as the aircraft have become older, so spare parts have become scarcer. To ease the burden, by 1988 the Flight's active strength had been reduced to four aircraft, the Swordfish, Firefly and two Sea Furies. The other aircraft were in safe storage, apart from the Sea Hawk, which was nearing the end of a major rebuild, to fly again successfully in the summer of 1989.

The problems of operating old aircraft made headline news in June 1989 when Lt.Cdr. John Beattie took off from Prestwick during the Scottish International Air Show. Only one undercarriage leg retracted due to a hydraulic fault. The Sea Fury is a little tricky to land, even on two wheels, and after trying every possible way to cure the problem, Beattie still had one wheel up and one down. He had no option but to fly out to sea and bale out. The Sea Fury was eventually recovered, but its plunge from 6,000 feet had smashed it irreparably. This is the only serious failure the Historic Flight have suffered in eighteen years of operating their special charges.

The pilots who fly these rare aeroplanes are all naval officers who have a "normal" job to do in the week, and the groundcrew also have other full time duties. Training and testing has to be fitted in whenever possible, and weekends are frequently sacrificed to show the Historic Flight's aircraft at air displays and special events. But neither pilots nor groundcrew begrudge the responsibility of maintaining these very special aircraft as a tribute to the Fleet Air Arm of the past.

101. Heron/Historic Flight in August 1972. The aircraft lined up are from front to rear: Hunter GA11, 4 Hunter T8s (for FRU/ADTU), Chipmunk (glider towing and air experience flights), 2 Sea Devons (VIP Transport); Tiger Moth, Firefly, Swordfish, Sea Fury (Historic aircraft); 2 Wessex Is (SAR Flight).

102. Firefly, Swordfish and Sea Fury, Historic Flight over Plymouth Hoe.

THE FLEET AIR ARM MUSEUM

For just over half of the station's active life Yeovilton has been the home of the Fleet Air Arm Museum. What began as a senior officer's bright idea became fact, and in its 25-year life the FAA Museum has grown to be one the major aviation museums of the world.

Until the Ilchester by-pass was built in the early 1970s the A303 trunk road from London to Exeter ran right alongside the air station. Yeovilton was about four hours drive from London, and many drivers stopped for a break to watch the flying. To prevent traffic hold-ups a small car park was created at the north east corner of the airfield in the early 1960s. At the time the station held three old aeroplanes, a Grumman Martlet, a Supermarine Seafire, and a Fairey Swordfish, and these three were parked alongside the viewing enclosure for the summer months. These minor attractions demonstrated that the general public were interested in aviation.

In the summer of 1962 Flag Officer Flying Training, Rear Admiral Percy Gick, visited his opposite number in Pensacola, Florida. He came away extremely impressed by US Naval Aviation's positive attitude to public relations, in marked contrast to his own service. He recommended the establishment of a Fleet Air Arm Museum and an enlarged viewing area for visitors. In October 1963 planning began for a Royal Review of the Fleet Air Arm, to be held at Yeovilton in May 1964 to celebrate the Silver Jubilee of naval control of the Fleet Air Arm. This gave the stimulus to the creation of a Museum. One of the original wartime hangars, No.11, was allocated, and an Engineer Officer, Cdr. Robin Foster was given the task of filling it by May. He bought some surplus display cases from the Science Museum for £100, and a letter from the Captain in the "Daily Telegraph" produced a flood of exhibits to fill them. Volunteers, including a number of station wives, helped to collect others and to set up a display. The three historic aircraft were wheeled in, three more were obtained from various sources and, on 29 May 1964, HRH The Duke of Edinburgh formally opened the Museum.

It remained open all summer, and attracted around 30,000 visitors. It closed for the winter but was ready again for the summer of 1965, by which time it was obvious that it was successful enough to be regarded as a permanent institution. This was achieved by the formation of a Charitable Trust to administer the Museum in 1966, and the first full time curator, Lt.Cdr. L.A."Harpy" Cox RN (Rtd) was appointed in 1967. He went on collecting aircraft and all the other memorabilia of naval aviation which were offered to him. By the early 1970s aircraft filled the original hangar and were overflowing onto the grass outside.

The trustees, at that time still all naval officers, were faced with a difficult decision. The trust had very little money, but if nothing was done, many important aircraft were just going to rot away. They took the courageous decision to appoint a professional fund raiser to find £250,000. Soon the building appeal was bringing in funds, and HRH The Prince of Wales took time off from his training with 707 Squadron to lay the foundation stone of the first new building. A Development Director, Cdr. Dennis White OBE, was brought in to oversee the building programme. At this time the Museum Director was usually the Air Engineering Officer of Yeovilton, but the museum was a full time job on its own. When Dennis White retired from the Navy in 1976 he became the first civilian Director. The constitution of the trustees was changed around this time, to include a number of civilians, but most of whom had served in the Fleet Air Arm. Their support and guidance was invaluable throughout this difficult and costly period of expansion.

The Caspar John Hall was opened in October 1975 by the Prime Minister, Mr Harold Wilson, in a ceremony hosted by Admiral of the Fleet Sir Caspar John, thus far the only naval aviator to become First Sea Lord and Chief

Early Wrens of all trades learned their jobs as they went along, but around the end of 1940 specialist training courses were introduced at Yeovilton for those Wrens working on aircraft and their associated equipment. One of the first courses was for Parachute Packers, an important job which paid 14 shillings per fortnight. However, at least one girl realised she could earn 21 shillings as a cook, and transferred to the galley.

There were no Wrens' quarters at Yeovilton in the early days. ''Immobile'' Wrens lived at home, and the others were billeted out in large houses in the surrounding countryside. Those girls who lived at Sparkford Rectory will always remember the mass of cockroaches which scurried from the kitchen when the light was turned on at night. The rectory was an old stone building, which was rarely warm. The bathroom was on the ground floor next to the larder, and condensation from the ceiling plopped back in icy drips upon the bather beneath. Depending on the size of the room, the girls slept in bunks up to eight in a room. The fire escape was a long ladder from one of the bedroom windows. It was never used in earnest but was a convenient way of sneaking in or out after hours.

Many Wrens came from naval families, and it was not surprising that they occasionally met their high ranking fathers in the course of their respective duties. Sheila Lyster was serving at Yeovilton at the time that her father, Rear Admiral Lumley Lyster, was inspecting the station. The Admiral stopped to speak to her and asked, ''What is your name?'' ''Lyster, Sir,'' she replied. ''Oh Yes,'' said the Admiral. ''I know your mother.''

Only the few WRNS officers were allowed into the wardroom, the rest of the Wrens were ratings and were subject to strict rules about not being seen in public with officers in uniform. The simple solution was to organise officers' dances away from the station, where the two groups could dance together as much as they liked.

There was one group of Wrens that had a particular influence on the training of Yeovilton's fighter pilots; the Cine-gun Assessors. The fighter aircraft used for mock

112. Sitting down job. W/T Wrens in Air Traffic Control.

113. Out and about job. Wren aerial photographers.

114. Highly trained job. 'Radio Wren'.

115. Wrens on Station Flight with the 'Admiral's Barge' in the 1950s

116. 'Met' Office in the 1960s

117. W/T Wren in the 1970s

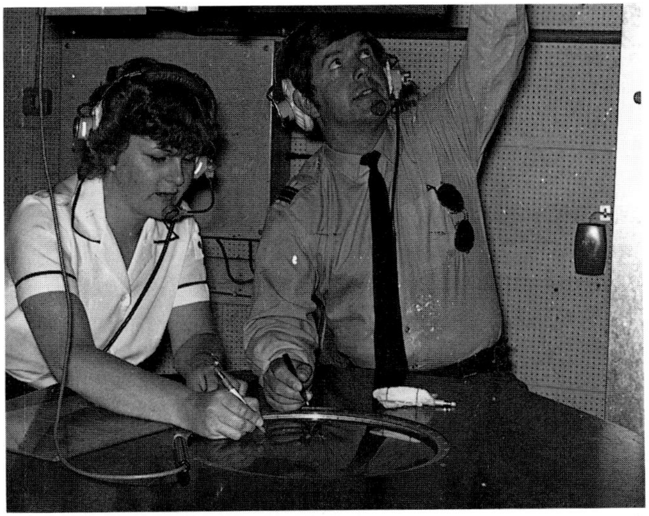

air combat were fitted with a Williamson G45 camera gun, aligned with the aeroplane's gun sight. When the pilot pressed the gun button, the camera filmed the position of the target aeroplane. When the film was developed after the sortie, the assessors ran the film through a projector and, with the help of various pre-computer aids, worked out the accuracy of the pilot's shooting. The instructors could then show the trainees just how good or bad their gunnery was.

Wrens working in the Fighter Direction School also made their contribution to the development of naval fighter tactics. In the early part of the war the Wrens only pedalled the tricycles used to simulate interception situations for trainee directors but, as time went on, Wrens became radio operators, and later trained as directors.

These were fairly clean tasks. The demand for trained air mechanics to serve with squadrons in carriers or overseas air stations led to a shortage of skilled men at home, and Wrens were recruited to replace them. A specialist training centre, HMS FLEDGELING, at Mill-meece in Staffordshire was opened in 1943 to train Wren Air Mechanics in the specialities of airframe, ordnance, engines, and electrical. Even more highly trained were the Wren Air Radio Mechanics. They spent more than a year going through both civilian technical college and practical service training before qualifying. By the end of the war, Yeovilton was just one of many FAA establish-ments which could not have operated without the efforts of very competent Wren mechanics. The girls replaced men in every sphere, and most of Yeovilton's aeroplanes were given their pre-flight daily inspections by Wren air mechanics.

Of course, the girls were subjected to all the practical jokes that apprentices traditionally suffer. A Wren nicknamed "Betty Ten", was part of the team trying to cure an engine which would not give full power. Her male section leader despatched her to the stores to fetch a tin of "rarified atmosphere", which he said might cure the problem. When Betty returned to the hangar there was no-one around, so to be helpful she opened the tin to remove its contents. She was horrified to find the tin was empty, and thought she must have let the "rarified atmosphere" escape. Betty shot back to the stores for another tin, only to be told that she had just collected the last one in stock.

Wrens continued to live in billets until the station's Wrens' Quarters were completed in 1943, at the far western end of the living site. Moral standards had to be maintained, and there was a separating wall between the men's blocks and the Wrens' huts. A very specific task of the night-time Duty Watch was to clear the "wailing wall" of loving couples at "pipe-down".

At its peak in the second world war the WRNS had nearly 75,000 members, of which 15,000 were working with the Fleet Air Arm, almost all of them in technical categories. With all the problems of dismantling the massive wartime navy and establishing peacetime requirements, the Admiralty did not confirm the continuing existence of the WRNS until 1949. There were still some 7,000 women in the service, and they could now look forward to working alongside the Royal Navy into the future.

Although Yeovilton has never been without Wrens, the girls' continuing support of the station's activities, was confined to a much more limited range of jobs than had been the case in wartime. Wren air mechanics were still an important part of the Fleet Air Arm, and at Yeovilton they formed much of the maintenance team of Station Flight, which was renamed Heron Flight in 1958. Heron Flight looked after the "Admiral's Barge" and the other communications aircraft, and also turned round visiting aircraft.

The flight prided itself on its ability to cope with any aircraft which came their way. In a typical year, 1966, they handled everything from a Stampe biplane to a Lockheed Starfighter, and from a Chipmunk to Shackletons. By 1972 a number of Hawker Hunters had been added, and Heron Flight was operating eight different types, totalling 15 aircraft in all. In August 1972 the Yeovilton Search and Rescue Flight was formed with two Wessex I helicopters, and the Wrens helped to service those too.

Despite the nightly clearing of the "wailing wall" it was inevitable that loving relationships would replace working ones in some cases. The station church of St.Augustine was granted a licence to perform marriages in the summer of 1968 and, on 1 June, P.O. John Squirrell and L/Wren Barbara Smith became the first couple to be married at Yeovilton. The captain lent his official car to take the newly-weds to their reception nearby. In August 1969 seven Wrens were married to seven sailors in the church, and the Daily Mirror carried the headline "Seven Brides for Seven Sailors".

No member of the WRNS has yet earned service "wings", but a small group at Yeovilton has earned something similar. When Heron Flight acquired its first Sea Heron as the Admiral's Barge in 1972, one of the flight's Wren air mechanics would usually travel as cabin attendant. They became particularly skilled as "air stewardesses" and were essential crew members on almost any trip in one of the Sea Herons. In 1977 a special (and unofficial) flying badge was introduced, and the first four received their "wings" from HRH Prince Charles.

Throughout its existence the Women's Royal Naval Service had been proud of being the only service which did not operate under a Discipline Act. Wrens were in fact "civilians in uniform", and members of the Royal Navy were not strictly bound to obey an order from a higher ranking member of the WRNS. This had never been a problem, but the arrangement did limit the range of jobs which Wrens, both officers and ratings, could take. In 1977 the WRNS gave up its unique status and was brought within the Naval Discipline Act, leading to much closer integration with the Royal Navy, and almost total inter-changeability of jobs. Wren air mechanics could now train as senior technical ratings. For the first time Wren air mechanics were allocated to front line squadrons, although they did not join their units when they embarked in the carriers. WRNS officers could also train to be Air Engineering Officers. A few of them had qualified at the end of the second world war and were waiting to take up their appointments when the war ended. In the late 1980s their daughters' generation were serving as AEOs on squadrons.

Although the Wrens' accommodation blocks at Yeovilton had been updated in the 1960s, they were still the same buildings put up during the second world war. New quarters were planned and built in the early 1980s, and were officially opened in late 1984 by HRH Princess Anne, who followed Princess Marina as only the second Chief Commandant in the history of the WRNS. "Never at Sea" was the WRNS' unofficial motto. The Boat's Crew Wrens of World War 2 had operated small launches in harbour, but members of the WRNS had never served in ships of the Royal Navy. In the new world of the 1980s Wren Air Mechanics not only joined front-line squadrons, but some of them actually served afloat on Royal Fleet Auxiliary ENGADINE during detachments from RNAS Culdrose.

The final barriers came down in February 1990, when it was announced in Parliament that members of the WRNS would serve at sea, fully integrated with the Royal Navy. Sitting at her rickety typewriter fifty years ago Wren Morton could never have imagined the variety of skills her contemporaries would deploy before their war ended. She would have dismissed as total fantasy a prediction that one day her "descendants" would carry out those duties in Her Majesty's Ships at sea.

118. The Supply Department provides clothing, and is staffed by Wrens.

119. Wren rating operating a lathe in the sheet metal repair section.

120. Working on an amplifier in the instrument repair shop.

Yeovilton, outpost of Westland.

Westland senior foreman Fred Walker first remembers Yeovilton on a July afternoon in 1940, when he was ordered to "Get on out to Ilchester" and meet Lord Aberconway who would be arriving to view the new site. Well, he knew where Ilchester was and eventually found Westland's new factory, just in time to see Lord Aberconway arrive with his entourage of Westland senior management, one of whom took him on one side and said, "We know you don't know anything about this, Walker, so if he asks any questions just let us do the talking."

Lord Aberconway, who in 1938 had replaced Sir Ernest Fetter as chairman of Westland, marched the length of the hangar and once round the paint shop, then without saying a word, got back into his car, and the party left. Fred was just wondering what to do next, when an aeroplane dropped out of the clouds and deposited its bombs on the nearby airfield. That was his introduction to what was to be his workplace for the next 12 years.

The Westland hangar at Yeovilton was set on the south side of the bend in Pyle Lane, about 400 ft. long by 100 ft. wide, with offices at the eastern end, overlooking the workshop floor. Large sliding doors along the hangar's south side, led out onto a concrete apron and aircraft park, from which a gateway gave access to the airfield on the other side of Pyle Lane. Here two further hangars were designated for Westland's use during the war, but later reverted to Yeovilton's Station Flight. One of them has since been demolished. A handful of experienced workers were transferred from Yeovil and Miss Young, the Women's Supervisor, was instructed to take on all the girls she could to supplement the work force. For training, every man had a mate, and the mate was a girl. That idea was popular.

Being on a naval air station, it's tempting to suppose that Westland were converting ex RAF aircraft into naval types, but in fact this was never done at Yeovilton. The majority of the work was rebuilding RAF Spitfires, interspersed with other urgent contracts on some of Westland's own marks. One of these contracts was to strip down Lysanders ready for crating and shipment to Malta. These were wanted so urgently, they were flown straight into Yeovilton and stripped down in the salvage hangar (on the other side of Pyle Lane. Any bad girls used to be sent over to the salvage department, it was a horrible job!

Another contract involving Lysanders was their conversion for secret operations in occupied Europe. This meant fitting a "Harrow" long range tank under the fuselage, and installing special radio equipment. The aircraft also had to be painted matt black, and a ladder fixed to the side. The ladder had to have a couple of luminous patches so that an agent waiting on the ground could find it easily in the dark, and climb aboard the slowly moving aircraft.

RAF pilots of Westland Whirlwind fighters would sometimes fly straight to Yeovilton for running repairs after being shot up. The Westland team would swing into action while the pilot waited, and as soon as the work was finished he would give it a quick test flight around the airfield, and if it was OK he would waggle his wings and disappear into the blue. On at least one occasion this practice aroused the wrath of the naval Commander (Flying), who complained it was setting a bad example for his trainee pilots.

Test flying of Westland's normal Yeovilton output was supposed to be done by Westland's own pilots, and they too occasionally fell foul of naval displeasure. Chief test pilot, Harald Penrose, certainly disliked being ordered about by the Navy, and would avoid working from Yeovilton if he could, leaving most of the Yeovilton flying to Jimmy Ramsden. In wartime the Yeovilton factory came under the jurisdiction of the Civilian Repair Organisation, and when the CRO phoned up one day to ask if there were any planes ready, they were told there

121. Interior of Westland's main hangar at Yeovilton in January 1942. Work is in progress on rebuilding Spitfires for the RAF. The offices at the far end have been rebuilt after the bomb damage of July 1940.

were several waiting for flying. "What do you mean, waiting for flying?" "Well, we can't get anybody up from Yeovil, they're all too busy." "Never mind your main works, you ring us when the planes are ready, and we'll send you pilots to fly them." Although this was contrary to Westland's contract, it was all eventually sorted out over a lunch at the Manor Hotel.

As the war came to an end Yeovilton's Westland hangar had very little work. Eventually the Air Ministry rang up the main works and said, "We must keep your Yeovilton unit going, and we're prepared, if you can send somebody up, to give you a refurbishing contract on 150 Spitfires." So, Fred Walker was despatched with Bill Samways, the chief inspector, around the squadrons to have a look at these Spitfires. At RAF Cosford, which was a "spit and polish" sort of station, they were shown a chart on the wall, showing all the stages of the job, totalling 180 hours. Having seen the aircraft, our Westland friends were a bit sceptical, but the Ministry said they were only prepared to pay for 180 hours on each Spitfire. The Spitfires were flown into Yeovilton, where Westland made the best of a bad job, but the first ones they delivered back to Cosford were rejected. According to Fred Walker, "There were loads of things wanted doing, but we didn't do them, because they only wanted us to do 180 hours work on them." What RAF Cosford had neglected to tell everyone about their 180 hour chart, was that if they found any damage they would take the plane out of the line, and it could be out for a month, while they were doing repairs on it, and then they would bring it back and carry on with the 180 hours. Finally the Ministry agreed, Westland could have the contract on a cost plus system.

By now the Spitfires were declining into second-line duties as jet fighters took over the air defence role. The Gloster Meteor had been the first jet to enter operational service in 1944, but the maximum operating height of the improved post-war versions was limited by pilot discomfort in the Meteor's unpressurised cockpit. Westland had gained valuable experience of pressurising cockpits during the development of the wartime high altitude fighter, the Westland Welkin, and the company were awarded a contract to modify the cockpits of about 30 Meteor F4s. The aircraft were modified in the Yeovilton hangar and this work, and the Spitfire contract, kept the operation going until 1952. The workforce had been reduced since the war years. Most of the female workers had already left, but now finding work became a real problem, with talk of making electric irons and the like. Time for the old hands to leave.

The parent company had by now made the move into helicopter construction, and a few Service Dragonflies and Whirlwinds trickled in to keep the refurbishment operations going. Another major contract for jet fighters was obtained in 1956.

Both the Gloster Meteor and its contemporary, the de Havilland Vampire, were 'straight winged' jets, a feature which limited their maximum speed. Warsaw Pact countries were already using the swept-wing MiG-15, which had outflown the Meteors it had met in combat over Korea. The North American Sabre (F86E) had proved equal to the MiG during the Korean War, and it was the only swept-wing fighter available to NATO forces for the Defence of Western Europe. Funded by the USA under the Mutual Defence Aid Programme, the RAF obtained 430 Sabres built by Canadair, delivered in 1952/3. When the Hawker Hunter finally became available in 1955/6, the Sabres were released by the RAF, who contracted with Westland for their refurbishment. The Sabres were turned out of Yeovilton in USAF markings, and test flown from Yeovilton or Merryfield before being handed back to the Americans. Most of them were then transferred under MDAP arrangements to other European countries, predominantly Italy and Yugoslavia.

By this time the company were wholly into helicopters, and their refurbishment, Dragonflies, Whirlwinds and latterly Wessex, kept Yeovilton's Hangar busy for many years. Many of Yeovilton's sailors were fascinated by the apparition 'parked' behind the hangar in the late 1960s, the Westland Westminster 'sky crane' helicopter, a project which never progressed beyond two prototypes.

Fixed-wing aircraft returned as a result of Westland having taken over Fairey Aviation in 1960, and the refurbishment of Fairey Gannets occupied much of the workforce. A Gannet was the last aircraft to leave the Yeovilton hangar in 1972, when the refurbishment operation was transferred to Weston-super-Mare. From then until 1985, when the hangar was purchased by the Royal Navy, Westland Yeovilton was a component repair facility, working on items such as wiring looms and hydraulic systems to support the main Yeovil factory. After a substantial rebuilding the hangar returned to housing Westland products, the Wessex and Sea King helicopters of 707 Squadron.

122. A Westland Lysander IIIA waits outside Westland's hangar at Yeovilton, after being painted matt black and fitted with a long range fuel tank, special radio equipment and a boarding ladder, ready for secret operations ferrying agents to and from occupied Europe.

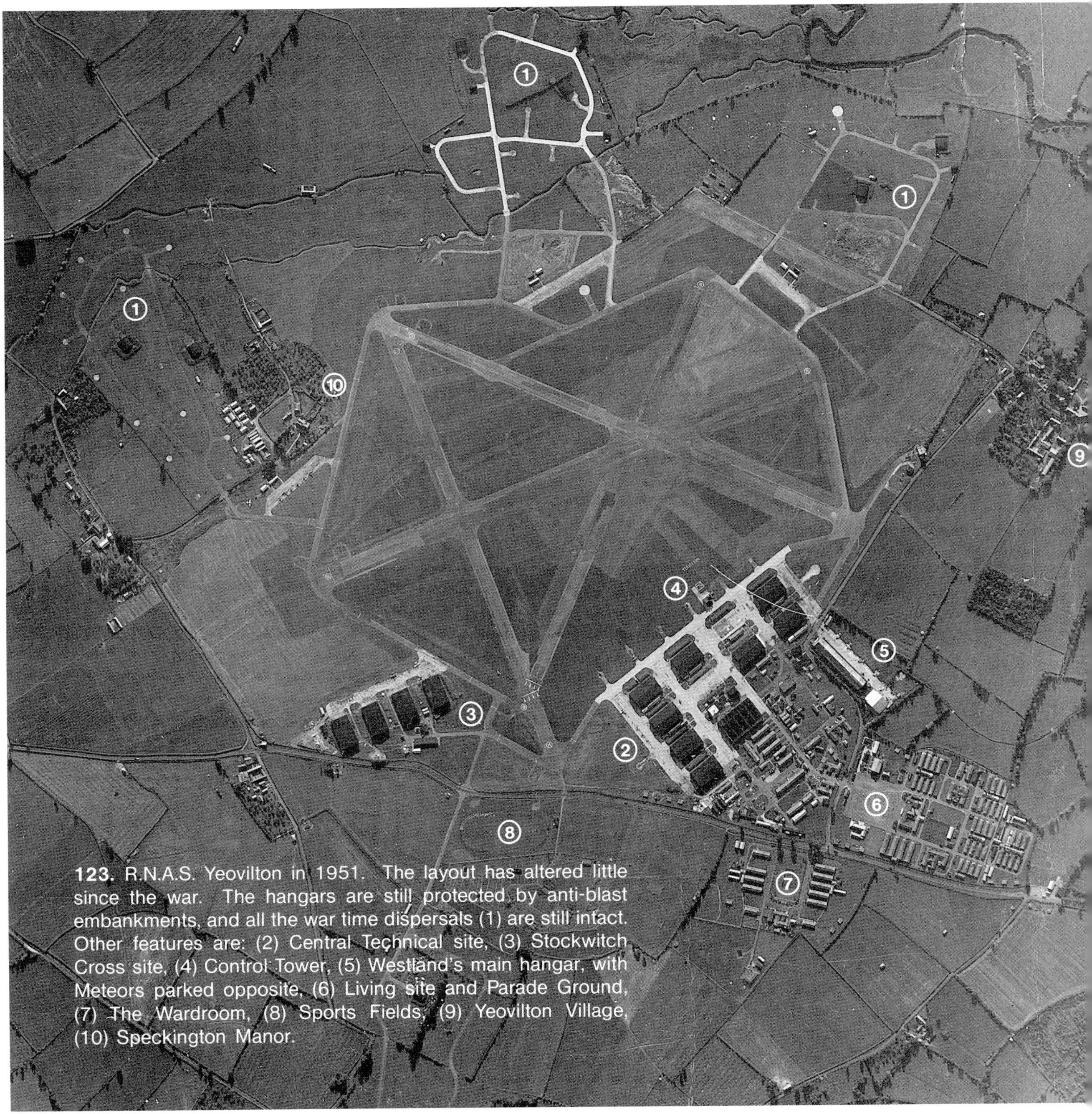

123. R.N.A.S. Yeovilton in 1951. The layout has altered little since the war. The hangars are still protected by anti-blast embankments, and all the war time dispersals (1) are still intact. Other features are: (2) Central Technical site, (3) Stockwitch Cross site, (4) Control Tower, (5) Westland's main hangar, with Meteors parked opposite, (6) Living site and Parade Ground, (7) The Wardroom, (8) Sports Fields, (9) Yeovilton Village, (10) Speckington Manor.

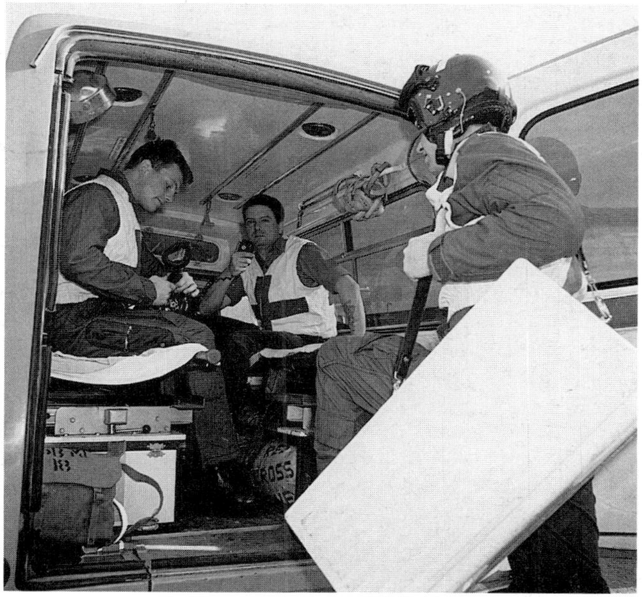

125. The emergency medical team.

124. Bird strikes can be a dangerous hazard, especially to low-flying fast jet aircraft. RNAS Yeovilton, one of the five busiest military airfields in Europe, is the headquarters of the Naval Air Command Bird Control Officer, with one of the most active and best equipped Bird Control Units in the UK. To keep the runways free of birds a variety of methods and equipment are used, including habitat management, shot guns, taped bird distress calls, "bird banger" pyrotechnics, and the controlled flying of birds of prey. The most important piece of equipment is a Landrover known as BABS (a mobile Bio-Acoustic Bird scaring System), crewed by a Falconer with his operational flight of four Peregrine or Lanner falcons, and a driver, who knows his way around the airfield like the back of his hand. It is a measure of their success that in an average year there is only one bird strike incident for every 125,000 aircraft movements.

126. Aircrew dinghy drills in the station's swimming pool.

127. The galley.

128. Armourer servicing a Sea Harrier's 30mm Aden gun.

129. The blacksmith shop.

130. Maintenance on a Pegasus engine in the engine repair department.

131. The air maintenance group carrying out deep maintenance on a Sea Harrier of 899 Squadron.

132. Chemical warfare exercise. The CO holds a muffled conference with his Nuclear Biological and Chemical Warfare Officer. On occasions everyone on the station dresses like this and tries to carry on as if everything is normal.

133. MEWSG (Maritime Electronic Warfare Support Group), is a NATO unit based at Yeovilton, headed by a Captain UK Navy, with a multi-national staff of about 40 military and civilian personnel. They operate three of these Transportable Radar and Communications Simulator Vans (TRACSVANs), which can be deployed either on land or on board ship. Each TRACSVAN has a three man crew, and is virtually self supporting, requiring only power and compass heading inputs from its host ship. When operated on land, a transportable generator set provides the power. Their task is "to provide a realistic hostile electromagnetic environment, in which NATO forces can conduct training, and exercise their units in all aspects of warfare". In 1987 MEWSG was selected by the International Electronic Warfare Association (Association of Old Crows) to receive their prestigious "Outstanding NATO Unit" award.

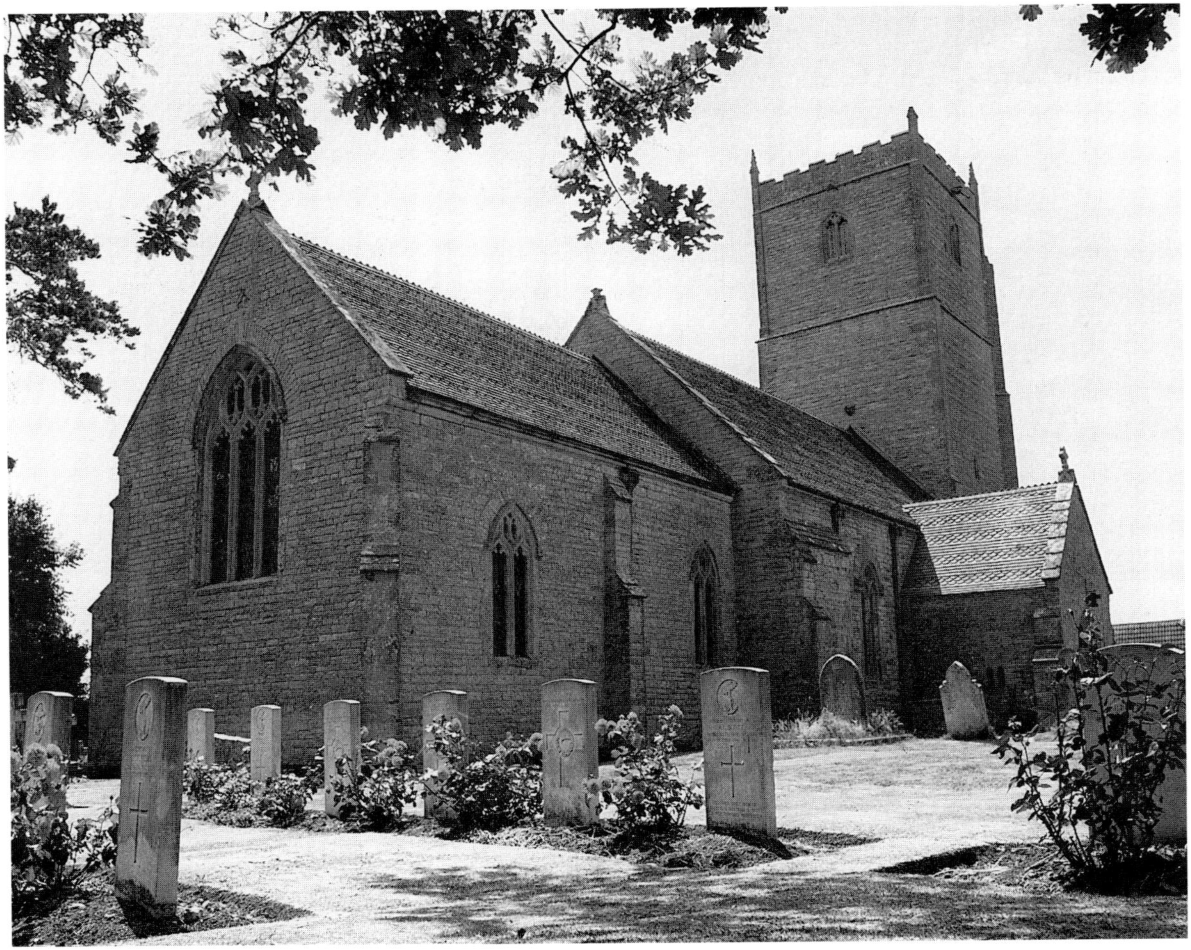

EPILOGUE

The Royal Naval Cemetery in the Churchyard of St Bartholomew's Church in Yeovilton is a poignant reminder that the 50th Anniversary of the Air Station has not been without human cost.

Interred in the Churchyard are the bodies of many men and women who died whilst serving in HMS HERON over the years. They rest in peace in this special place that was their life in War and Peace.

In the Memorial Chapel in St Augustine's Church rest the Fleet Air Arm Book of Remembrance and other Books of Remembrance from Aircraft Carriers and Naval Air Stations that have now paid off.

In this Anniversary Year it is hoped that The Fleet Air Arm will be able to purchase and restore St Bartholomew's Church Yeovilton, and that it will become the Fleet Air Arm Memorial Church.

GLOSSARY

ADR - Air Direction Radar
ADTU - Air Direction Training Unit
AEW - Airborne Early Warning
AFDU - Air Fighting Development Unit
AI - Air Interception (usually describing an airborne radar)
ASW - Anti Submarine Warfare
bowser - motorised petrol tanker for refuelling aircraft anywhere on the station
Boyd Trophy - Silver model of Fairey Swordfish, named in honour of Admiral Sir Dennis Boyd; awarded annually to the aircrew who have achieved the finest feat of naval aviation in that year
CAM Ship - Catapult Armed Merchant Ship
CAP - Combat Air Patrol
"Clockwork" - NATO exercise in arctic conditions in Norway
CO - Commanding Officer
DFC - Distinguished Flying Cross
DSC - Distinguished Service Cross
DSO - Distinguished Service Order
ECM - Electronic Counter Measures
FAA - Fleet Air Arm
FAAM - Fleet Air Arm Museum
FAB - Forward Air Base (for servicing, refuelling and rearming aircraft in the field)
FAW - Fighter All Weather
FB - Fighter Bomber
FDO - Fighter Direction Officer
FOB - Forward Operation Base (nearer the battlefield than the FAB, for refuelling and rearming only)
FOFT - Flag Officer Flying Training (which later became FONFT)
FONAC - Flag Officer Naval Air Command (an amalgamation of the duties of FONFT and Flag Officer Air Home, in 1970)
FONFT - Flag Officer Naval Flying Training
FRADU - Fleet Requirements and Air Direction Unit, an amalgamation of FRU and ADTU
FRU - Fleet Requirements Unit, providing aircraft for tasks such as target towing and radar calibration, in support of Fleet and FAA training programmes
HAS - Helicopter Anti-Submarine
HC - Helicopter Commando
HMS Dipper - RNAS Henstridge
HMS Humming Bird - RNAS Zeals
HU - Helicopter Utility
IFTU - Intensive Flying Trials Unit
Inskip Award - The recommendation by Sir Thomas Inskip, Minister for the Coordination of Defence, in July 1937, which transferred control of naval aviation from the RAF to the Royal Navy
INTU - Indian Navy Training Unit
MEWSG - Maritime Electronic Warfare Support Group (a NATO unit simulating communications jamming for training purposes)
M/T or MT - Motor Transport
NACSSU - Naval Aircraft Squadron Support Unit
NAFDU - Naval Air Fighting Development Unit
NAFS - Naval Air Fighter School
NFSF - Naval Flying Standards Flight, nicknamed "the trappers"
RAN - Royal Australian Navy
Red Dragon Flight - formed within 707 Squadron, specifically to include the Prince of Wales whilst training on Wessex helicopters
RFA - Royal Fleet Auxiliary
RNAS - Royal Naval Air Station
RWP - Runway Working Party
SBAC - Society of British Aerospace Constructors
VC - Victoria Cross

RANKS

Admiral of the Fleet
Adm. - Admiral
V.Adm. - Vice Admiral
R.Adm. - Rear Admiral
Cdre. - Commodore
Capt. - Captain
Cdr. - Commander
Lt.Cdr. - Lieutenant Commander
Lt. - Lieutenant
S/Lt. - Sub Lieutenant
Mid. - Midshipman
(E) - denotes an engineer officer when suffixed to the rank, e.g. Lt.(E).
Surg. - denotes a Medical Officer when prefixed to the rank, e.g. Surg.Cdr.

COMMANDING OFFICERS - H.M.S. HERON

Captain H.S. Murray-Smith, RN	18.6.40 to 5.6.42
Captain M.S. Thomas, DSO, RN	5.6.42 to 2.9.43
Captain H.M.C. Abel Smith, RN	2.9.43 to 23.5.44
Captain C.L. Keighley-Peach, DSO, OBE, RN	23.5.44 to 18.8.45
Captain J.B. Heath, OBE, RN	18.8.45 to 12.2.47
Captain W.K. Edden, OBE, RN	12.2.47 to 30.5.49
Captain W.A. Adair, OBE, RN	30.5.49 to 5.1.51
Captain R.T. Paul, CBE, RN	5.1.51 to 1.5.52
Captain E.V.St.J. Morgan, DSC, RN	1.5.52 to 8.9.53
Captain A.F. Black, DSC, RN	9.9.53 to 24.7.55
Captain H.J.F. Lane, OBE, RN	25.7.55 to 16.1.58
Captain D.B. Law, MBE, DSC, AR.Ac.S., RN	17.1.58 to 11.1.60
Captain W.C. Simpson, OBE, DSC, RN	11.1.60 to 8.1.62
Captain R.H.P. Carver, CBE, DSC, RN	8.1.62 to 20.2.64
Captain T.G.V. Percy, OBE, DSC, RN	21.2.64 to 31.3.66
Captain J.D. Honywill, RN	1.4.66 to 10.7.66
Captain G.C. Baldwin, DSC, RN	11.6.66 to 3.5.68
Captain C.J. Cunningham, DSC, RN	3.5.68 to 19.5.70
Captain A.R. Rawbone, AFC, RN	19.5.70 to 3.2.72
Captain K.A. Leppard, RN	4.2.72 to 22.2.74
Captain J.J. Phillips, RN	22.2.74 to 21.11.75
Captain H.J. Abraham, RN	21.11.75 to 27.10.77
Captain B.J. Williams, RN	27.10.77 to 5.8.80
Captain M.J.F. Rawlinson, RN	5.8.80 to 26.2.82
Captain P.J. Williams, OBE, RN	26.2.82 to 13.1.84
Captain R.J. Northard, RN	13.1.84 to 4.2.86
Captain R.P. Warwick, RN	4.2.86 to 9.2.88
Captain C.L. MacGregor, RN	9.2.88 to 10.10.89
Captain R. Shercliff, RN	10.10.89 to present

PRINCIPAL TRAINING SQUADRONS AT YEOVILTON AND SATELLITES 1940–1990

SQUADRON No.	ROLE	PRINCIPAL AIRCRAFT	AIRFIELD* and DATES	SUBSEQUENT POSTING
700	a) Maintenance Test Pilot Training b) Trials Unit	a) Firefly, Oxford, Seafire, Sea Otter, Avenger, Barracuda, Sea Fury b) Sea Hawk, Sea Venom, Whirlwind	a) Y 4/46 to 9/49 b) Y&M 9/58 to 7/61	a) Disbanded b) Disbanded
700A	Sea Harrier IFTU	Sea Harrier FRS1	Y 6/79 to 3/80	Into 899 Sqdn
700L	Lynx IFTU	Lynx HAS2	Y 9/76 to 12/77	Into 702 Sqdn
700P	Phantom IFTU	Phantom FG1	Y 4/68 to 3/69	Into 767 & 892 Sqdns
700Y	Sea Vixen IFTU	Sea Vixen FAW1	Y 11/58 to 7/59	Into 892 Sqdn
702	Lynx HQ and Training Sqdn	Lynx HAS2 & 3	Y 1/78 to 7/82	Portland
707	Command Helicopter Advanced and Operational Flying Training	Wessex HU5, Sea King HC4	Y 5/72 to Present	
736	School of Air Combat	Seafire, Barracuda	Y 5/43 to 9/43	St. Merryn
748	No. 10 Naval Operational Training Unit	Martlet, Reliant, Seafire, Corsair, Hellcat, Firefly	Y 3/44 to 10/44	Dale
750	No. 1 Observers Training School	Shark	Y 5/40 to 10/40	Trinidad
751	No. 1 Observers Training School	Walrus	Y 5/40 to 8/40	Arbroath
752	No. 1 Observers Training School	Proctor, Vega Gull, Albacore	Y 5/40 to 10/40	Trinidad
759	Fleet Fighter School	Skua, Fulmar, Sea Hurricane, Seafire, Wildcat, Corsair, Hellcat, Firefly	Y 9/40 to 9/45 Z 9/45 to 1/46	Disbanded
760	a) Fighter Pool b) Corsair Familiarisation	a) Skua, Fulmar, Sea Hurricane, Martlet b) Corsair, Harvard	a) Y 9/40 to 12/42 b) Z 4/45 to 9/45	a) Disbanded b) Disbanded
761	Advanced Training	Fulmar, Sea Hurricane	Y 8/41 to 7/42	Disbanded
764	Operational Flying School	Seafire, Firefly, Sea Hawk	Y 9/53 to 11/54	Disbanded
766	All Weather Fighter Pool/School	Sea Venom, Sea Vixen FAW1, Sea Vixen FAW2	Y 10/55 to 11/56 M 11/56 to 1/58 Y 1/58 to 12/70	Disbanded
767	a) Deck Landing Control Officer Training ("Clockwork Mouse Sqdn.") b) Phantom Conversion Unit	a) Seafire, Firefly, Sea Fury b) Phantom FG1	a) Y 9/49 to 1/52 M 1/52 to 9/52 b) Y 1/69 to 8/72	a) Stretton b) Disbanded
780	Pilot Training	Tiger Moth, Miles Master, Harvard, Swordfish, Blenheim	CH 10/43 to 11/44	Lee on Solent
787	Fleet Fighter Development	Fulmar, Sea Hurricane, Martlet	Y 3/41 to 6/41	Duxford
790	Fighter Direction School	Oxford, Fulmar, Seafire	CH 7/42 to 4/45 Z 4/45 to 9/45	Dale
794	a) Air Target Towing Unit b) No. 1 NAFU	a) Roc, Swordfish, Skua b) Sea Hurricane, Martlet, Master, Defiant	a) Y 8/40 to 7/43 b) CH 12/43 to 6/44	a) Angle b) Disbanded
799X	Flying Check and Conversion Refresher Unit	Harvard, Firebrand, Sea Fury	Y 5/48 to 12/51	Machrihanish

*Y – Yeovilton, Z – Zeals, CH – Charlton Horethorne, M – Merryfield

FRONT LINE SQUADRON AT YEOVILTON AND SATELLITES 1940–1990

SQUADRON	AIRCRAFT	AIRFIELD* and DATES	TASK AT AIRFIELD	SUBSEQUENT EMBARKATION and OPERATIONAL AREA
800 Fleet Fighter	Fulmar II	Y 30/6/41 to 12/7/41	Work Up	FURIOUS, INDOMITABLE – Malta and Arctic Convoys, W. Indies
Fighter Reconnaisance	Sea Harrier (4)	Y 3/80 to Present	(Parent Station)	INVINCIBLE, HERMES, ILLUSTRIOUS
801 Fleet Fighter	Sea Hurricane I(a) & I(b) (12)	Y 1/8/41 to 6/10/41	Reform and Work Up	ARGUS, EAGLE – Mediterranean
Fighter Reconnaisance	Sea Harrier FRS1 (5)	Y 1/81 to Present	(Parent Station)	INVINCIBLE
802 Fleet Fighter	Sea Hurricane I(b) & I(c) (6)	Y 1/2/42 to 17/4/42	Reform and Work Up	AVENGER – Arctic Convoys, N. African Landings, Sqdn lost when Avenger was torpedoed 15/11/42 during 'Torch'
	Sea Fury FB11, T20 (13)	Y 12/48 to 1/49 4/50 to 7/50	Work Up Work Up & Air Display	VENGEANCE
	Sea Hawk FB5 (11)	M 1958	Transit	ARK ROYAL
803 Fleet Fighter	Scimitar F1	Y 5/59 and 9/59	Transit	VICTORIOUS
804 Cam Ship Fighters	Fulmar I, II Sea Hurricane I(b), II(b)	Y 10/2/41	Detachments to Cam Ships	RAF Sydenham and Detachments to PEGASUS, ARIGUANI, MAPLIN, SPRINGBANK, SS MICHAEL E, – Atlantic Convoys
Fleet Fighter		Y 14/5/42	Work Up	ARGUS, FURIOUS
	Sea Hurricane IIc	CH 6/4/43 to 20/6/43	Waiting to Re-Equip with Hellcat Is	RAVAGER, EMPEROR – USA-Bound Convoys
806 Fighter	Seafire F XV (12)	Y 2/46 to 4/46	Work Up	GLORY
	Sea Hawk F1, FB3 (12)	Y 6/54 to7/54	Work Up	CENTAUR
807 Fleet Fighter	Fulmar I (12)	Y 9/12/40 to 4/2/41	Work Up	FURIOUS, ARK ROYAL – Atlantic and Malta Convoys
	Seafire I(b), II(c)	Y 12/7/42 to 24/8/42	Re-Equip and Work Up (First Sqdn with Seafires)	FURIOUS – N. African Landings
	Sea Fury FB11 (12)	Y 3/49 to 5/49	Re-Equip and Work Up	THESEUS
808 Fleet Fighter	Fulmar II (6)	Y 18/4/42 to 30/5/42	Work Up	BITER – Atlantic
	Seafire L II(c) (9)	CH 11/12/42 to 8/1/43	Work Up	BATTLER – Gibraltar Convoy
	Seafire L II(c) (9)	Y 20/5/43 to 4/6/43	Work Up	BATTLER – Salerno Landings
Naval Gunfire Support and Reconnaisance	Spitfire V(b) LF V(b) Seafire L II(c)	H 7/3/44 to 22/4/44	Work up with 3rd Naval Fighter Wing	'D' Day Landings, Normandy.
808 Royal Australian Navy	Sea Venom FAW20, 53 (8)	Y 8/55 to 2/56	Reformed as RAN Sqdn Re-Equip and Work Up	HMAS MELBOURNE
809 All Weather Fighter	Sea Venom FAW20 (9) FAW21	Y 5/54 to 1/2/55 25/2/55 to 20/3/56	Reformed and Re-Equip Re-Equip	BULWARK, ARK ROYAL, ALBION – Suez Malta (Disbanded at Yeovilton 1956)
	Sea Venom FAW21 (9)	Y&M 3/57 to 8/59	Work Up	ALBION
Strike	Buccaneer S2 (8)	Y 2/9/68 to 23/9/68 Y 9/7 70 to 25/7/70	Exercises Exercises and Air Displays	Returned to Lossiemouth
All Weather Fighter	Sea Harrier FRS1 (8)	Y 4/82 to 12/82	Re-Commissioned	MV ATLANTIC CONVEYOR, HERMES, INVINCIBLE - South Atlantic

*Y – Yeovilton, Z – Zeals, CH – Charlton Horethorne, M – Merryfield

FRONT LINE SQUADRON AT YEOVILTON AND SATELLITES 1940-1990

SQUADRON	AIRCRAFT	AIRFIELD* and DATES	TASK AT AIRFIELD	SUBSEQUENT EMBARKATION and OPERATIONAL AREA
810 Fleet Fighter	Firefly FR4 (12)	Y 3/49 to 5/49	Work Up	THESEUS
811 Fleet Fighter	Sea Hawk FB3 (10)	Y 8/55	Exercises and Battle of Britain Fly Past	BULWARK, ALBION
814 Fleet Fighter	Firefly FR4 (8)	Y 12/48 to 1/49	Transit	VENGEANCE
Anti-Submarine	Sea King HAS1 (10)	Y 21/6/77 to 28/6/77	Transit	HERMES
815 Anti-Submarine	Lynx HAS2	Y 1/81 to 7/82	H/Q Squadron	Various Ship's Flights
819 Anti-Submarine	Wessex HAS1 (2)	Y 5/64	Detachment	RFA TIDEPOOL
827 Torpedo Spotter Reconnaisance	Albacore I (12)	Y 15/9/40 to 2/11/40	Formation	RAF Coastal Command off Scotland
831A Electronic Warfare	Sea Venom 22 ECM (3)	Y 10/65	Detachment	Various Fleet Carriers
845	Wessex HAS1 (12)	Y 9/62	Exercise	ALBION
	Wessex HAS1, HU5 (14)	Y 9/69	Detachment	
Commando	Wessex HU5, Sea King HC4 (12)	Y 9/72 to Present	(Parent Station)	BULWARK, ALBION, INTREPID, HERMES, FEARLESS, Several RFAs
846 Commando	Wessex HU5 (4)	Y 5/72 to 12/75	(Parent Station)	RFAs, REGENT, RESOURCE, TIDESPRING
	Wessex HU5 (8)	Y 1/4/76 to 5/9/77	(Parent Station)	HERMES
	Wessex HU5 (NACSSU)	Y 11/77 to 10/78	(Parent Station)	
	Wessex HU5, Sea King HC4	Y 10/78 to Present	(Parent Station)	BULWARK, HERMES, INTREPID, FEARLESS, SS CANBERRA, MV NORLAND - S. Atlantic
848 Cammando	Wessex HU5 (16)	Y 12/72 to 3/76	(Parent Station)	BULWARK
	Wessex HU5 (13)	Y 4/82 to 11/82	(Parent Station)	SS ATLANTIC CONVEYOR, RFA's REGENT, OLNA, FORT AUSTIN, HMS ILLUSTRIOUS - S. Atlantic
880 Fleet Fighter	Sea Hurricane I(b) (9)	Y 17/7/41 to 20/7/41	Work Up	FURIOUS, INDOMITABLE Arctic Convoys, West Indies
	Sea Hurricane I(b) (9)	Y 5/10/41 to 10/10/41	Work Up	
882 Fleet Fighter (& Escort Carrier)	Sea Hurricane I(b) Martlet I (9)	Y 1/12/41 to 15/12/41	Work Up	ILLUSTRIOUS - Madagascar ARCHER - Eastern Seaboard USA
883 Fleet Fighter	Sea Hurricane I(b), I(c) (6)	Y 10/10/41 to 28/1/41	Formation and Work Up	AVENGER - Arctic Convoys
884 Fleet Fighter	Fulmar II (6)	Y 7/2/42 to 22/3/42	Work Up	VICTORIOUS - Malta Convoy
885 Fleet Fighter	Sea Hurricane I(b)	Y 1/12/41 to 14/3/42	Reform and Work Up	VICTORIOUS - N. African Landings
Naval Gunfire Support and Reconnaisance	Seafire L III, F III	H 31/3/44 to 22/4/44	Work up with 3rd Naval Fighter Wing	'D' Day Landings, Normandy
886 Fleet Fighter	Fulmar II (6)	Y 22/6/42 to 10/7/42 CH 10/7/42 to 11/8/42	Work Up	RAF Fighter Command, Scotland
Naval Gunfire Support and Reconnaisance	Seafire L III (10)	H 11/3/44 to 25/4/44	Work Up with 3rd Naval Fighter Wing	'D' Day Landings, Normandy
887 Fleet Fighter	Fulmar II (6)	Y 1/6/42 to 10/7/42 CH 10/7/42 to 25/7/42	Work Up	N. Ireland
	Seafire F III, L III	H 13/12/43 to 8/1/44	Work Up with 24th Naval Fighter Wing	INDEFATIGABLE - off Norway, Ceylon

FRONT LINE SQUADRON AT YEOVILTON AND SATELLITES 1940-1990

SQUADRON	AIRCRAFT	AIRFIELD* and DATES	TASK AT AIRFIELD	SUBSEQUENT EMBARKATION and OPERATIONAL AREA
890 All Weather Fighter	Sea Venom FAW20 (9)	Y 3/54 to 7/55	Reformed as AW Sqdn (Parent Station)	BULWARK, ALBION
	Sea Venom FAW21 (9)	Y 2/56 to 6/56	(Parent Station)	ALBION
	Sea Vixen FAW1 (10)	Y 2/60 to 10/66	(Parent Station)	HERMES, ARK ROYAL
	Sea Vixen FAW 2 (6)	Y 8/67 to 8/71	(Parent Station)	
891 Fleet Fighter	Sea Hurricane I(b) (6)	CH 11/8/42 to 7/9/42	Work Up	ARGUS, DASHER - N. African Landings
All Weather Fighter	Sea Venom FAW20 (9)	Y 11/54 to 9/56	(Parent Station)	ARK ROYAL
	Sea Venom FAW21, 22 (8)	Y&M 11/56 to 7/61	(Parent Station)	ARK ROYAL, BULWARK, CENTAUR
891X Training	Sea Venom FAW20 Vampire FB5	Y 3/55 to 8/55	RAN Crew Training prior to Formation of 808 Sqdn	
892 All Weather Fighter	Sea Venom FAW21 (8)	Y 7/55 to 7/56	Reformed as AW Sqdn (Parent Station)	ALBION, EAGLE - Suez
	Sea Vixen FAW1, 2 (12)	Y 7/59 to 10/68	(Parent Station)	ARK ROYAL, VICTORIOUS, HERMES, CENTAUR
	Phantom FG1	Y 3/69 to 7/72	(Parent Station)	ARK ROYAL
893 Fleet Fighter	Martlet I, II & IV (10)	CH 9/9/42 to 6/10/42	Work Up	FORMIDABLE - N. African Landings
	Martlet IV	Y 16/11/43	Disbanded	
All Weather Fighter	Sea Venom FAW21, 22 (9)	Y&M 2/56 to 2/60	Reformed as AW Sqdn	BULWARK, EAGLE - Suez ARK ROYAL, VICTORIOUS
	Sea Vixen FAW1 (6)	Y 9/60 to 11/65	Reformed	ARK ROYAL, CENTAUR, VICTORIOUS
	Sea Vixen FAW2 (11)	Y 11/65 tp 7/70	Reformed	VICTORIOUS, HERMES
894 Fleet Fighter	Seafire F III (12)	H 19/10/43 to 8/1/44	Work Up with 24th Naval Fighter Wing	INDEFATIGABLE - off Norway, Ceylon
All Weather Fighter	Sea Venom FAW21, 22 (12)	Y&M 1/57 to 12/60	(Parent Station)	EAGLE, VICTORIOUS, ALBION
897 Fleet Fighter	Sea Hurricane I(b) Replaced by Seafire I(b) (10)	CH 11/1/43 to 22/3/43	Re-Equip and Work Up	UNICORN - Salerno Landings
Naval Gunfire Support and Reconnaisance	Spitfire LF V(b)	H 2/3/44 to 6/5/44	Work Up with 3rd Naval Fighter Wing	'D' Day Landings - Normandy
899 FAW Fighter	Sea Vixen FAW1	Y 2/61 to 6/64	Reformed as HQ Sqdn	ARK ROYAL, HERMES, CENTAUR, EAGLE, VICTORIOUS
	Sea Vixen FAW2	Y 6/64 to 1/72	Reformed as IFTU	EAGLE
	Sea Harrier FRS1, T4N	Y 3/80 to Present	Reformed as IFTU & HQ Sqdn	HERMES, INVINCIBLE, ILLUSTRIOUS
1770 Two Seat Fleet Fighter	Firefly I (12)	Y 10/9/43 to 14/12/43	Formation and work Up (First Sqdn to have Fireflies)	INDEFATIGABLE - Off Norway, Ceylon
1771 Two Seat Fleet Fighter	Firefly I (12)	Y 1/2/44 to 3/3/44	Formation and Work Up	TRUMPETER, RAVAGER (Deck Landing Trainining) IMPLACABLE - Off Norway, Far East
1834 RNVR Training	Sea Fury FB11, T20	Y 1/54 to 4/55	Southern Air Division RNVR	Disbanded
3 Commando Brigade Air Sqdn	Gazelle, Sioux, Lynx	Y 1982 to Present	Commando Support	

Back Cover: A view across the lift shaft of ARK ROYAL IV in the late 1970s. The Phantoms of 892 Squadron and the Buccaneers of 809 Squadron epitomise the role of the Fleet Air Arm - aircraft on ships, ready to deliver an effective presence anywhere in the world.